Book Title: Analysis of Radio-Propagation Environments to Support Standards Development for RF-Based Electronic Safety Equipment

Book Author: Catherine A. Remley; William F. Young; Jacob L. Healy;

Book Abstract: We analyze data from NIST field tests in which radio-propagation channel characteristics were measured at approximately the same physical locations where the performance of various RF-based firefighter distress beacons was tested. These side-by-side tests were made in representative emergency responder environments, including an apartment building, four types of office buildings, a convention center, and an urban canyon. These environments contain propagation features that often impair radio communications, including stairwells, basements, and rooms deep within buildings, among others. The goal of this work is to determine appropriate performance metrics for use in the development of laboratory-based test methods for RF-based electronic safety equipment. For the structures we studied, we found that attenuation, rather than multipath, plays a more significant role in determining whether or not a remote distress alarm is received outside the structure. The analysis has enabled rough classification of structures into categories of attenuation values that can be used in laboratory-based test methods to verify the performance of the RF-based alarm system that we tested. The environments, tests, and measured data are discussed in detail.

Citation: NIST TN - 1559

Keywords: attenuation; delay spread; emergency responders; firefighter communications; multipath; public-safety radio communications; radio propagation experiments; transfer function; urban canyon; wireless communications

National Institute of
Standards and Technology
U.S. Department of Commerce

NIST Technical Note 1559

Analysis of Radio-Propagation Environments to Support Standards Development for RF-Based Electronic Safety Equipment

Kate A. Remley
William F. Young
Jacob Healy

NIST Technical Note 1559

Analysis of Radio-Propagation Environments to Support Standards Development for RF-Based Electronic Safety Equipment

Kate A. Remley
William F. Young
Jacob Healy

Electromagnetics Division
National Institute of Standards and Technology
325 Broadway
Boulder, CO 80305

U.S. Department of Commerce
John Bryson, Secretary

National Institute of Standards and Technology
Patrick D. Gallagher, Director

National Institute of Standards and Technology Technical Note 1559
Natl. Inst. Stand. Technol. Tech. Note 1559, 61 pages (March 2012)
CODEN: NTNOEF

U.S. Government Printing Office
Washington: 2005

For sale by the Superintendent of Documents, U.S. Government Printing Office
Internet bookstore: gpo.gov   Phone: 202-512-1800   Fax: 202-512-2250
Mail: Stop SSOP, Washington, DC 20402-0001

Contents

Executive Summary

The National Institute of Standards and Technology (NIST) has been involved in a multi-year project to support the development of performance metrics and test methods for radio-frequency (RF)-based electronic safety equipment used by the public-safety community. The work reported here focuses on side-by-side measurements of radio-propagation environment characteristics and actual wireless-device performance in representative emergency responder environments. Identifying the radio-channel characteristics that significantly impair wireless-device performance in various environments enables the development of standardized laboratory-based test methods that simulate the conditions under which electronic safety equipment will be used in the field. The test methods can then be incorporated into consensus standards for this equipment.

The analysis presented here has been funded by the U.S. Department of Homeland Security's Standards Branch. The work reported here has focused on RF-based personal alert safety systems (PASS), used by firefighters to indicate when a firefighter is motionless or in distress. However, the methodology and analysis presented here could easily be extended to apply to other types of wireless devices that operate in a point-to-point mode.

In the propagation-channel studies, NIST engineers measured path loss ("attenuation") and the level of reflectivity (or "multipath," here quantified by the root-mean-square delay spread) in large public structures and environments where radio communications could be difficult. These environments include multi-story buildings; buildings with subterranean floors and tunnels; buildings with deep interior spaces; those with few windows; and outdoor "urban canyons," consisting of city streets surrounded by tall buildings. The NIST Public-Safety Communications Research Lab has funded the measurements of the propagation channel.

To support development of standards in public-safety applications, the NIST studies focused on the penetration of radio signals from outside to inside a given structure (and vice versa), as opposed to outdoor-to-outdoor or within-building tests. To simulate an incident command post, in the propagation-channel studies a transmit antenna was positioned outside of each structure at a location representative of a fireground configuration. The receive antenna was then placed at various discrete locations within the environment at progressively greater distances from the transmit antenna. At each location, the path loss and RMS delay spread were measured with a vector-network-analyzer-based measurement method that is described in Section 2. NIST propagation-channel tests have been conducted in several environments, seven of which are described here.

RF-based PASS measurements were conducted in approximately the same locations as the channel characterization measurements. The base-station unit was placed at the location of the VNA transmit antenna. Portable RF-based PASS devices were then carried to approximately the same locations where the receive antenna had been located. At each test location, the operator of the portable PASS unit activated an RF-based alarm, and the base-station operator noted whether the alarm was received, not received, or received after a significant delay.

1

Even though RF-based PASS systems are capable of two-way communications, the NIST tests focused on whether or not the alarm from the portable RF-based PASS device was received by the base station, rather than whether the portable device received an alarm signal from the base station. This is because the portable RF-based PASS device generally transmits with a lower power than does the base station in order to conserve battery life and, in some cases, to meet "intrinsic safety" standards for electronic equipment. Consequently, the signal emitted from the RF-based PASS device is typically weaker, and testing for reception of the alarm signal by the base station represented a worse-case scenario.

We tested two different, commercially available, RF-based PASS systems, one that operates on a licensed frequency in the 450 MHz public-safety narrowband frequency allocation, and one that operates in the unlicensed spectrum between 902 MHz and 928 MHz. The latter system was tested alone and with one repeater unit. At the time the tests were conducted, these were the only two RF-based PASS systems that were commercially available in the U.S. In the meantime, other manufacturers have implemented RF-based PASS, and we expect that use of this technology will continue to increase.

From our analysis of the side-by-side measurements described here of the radio-propagation channel and RF-based PASS devices, we have been able to draw several important conclusions. First, the data indicate that attenuation, rather than multipath, is the most common cause of a missed alarm for an RF-based PASS device in the representative medium-to-large-structure, radio-propagation environments that were studied.

A second conclusion is that there is a range of path-loss values that can be used to roughly classify various structures as low-, medium-, and high-attenuation environments. In low-attenuation environments (defined here as less than 100 dB path loss at a frequency of approximately 750 MHz), RF-based PASS systems could typically be operated successfully without a repeater. In the medium-attenuation environments (approximately 100 dB to 150 dB of path loss), RF-based PASS systems could typically be operated with a single repeater. In the high-attenuation environments (path loss greater than 150 dB), we expect that RF-based PASS systems may encounter difficulty with efficient and reliable RF transmission (using 2010 technology consisting of a base-station transceiver and portable, body-worn transceivers).

A third conclusion can be drawn by looking at the RF-based PASS performance tests conducted in an apartment building with a cell-phone base station located on its roof. This structure did not present significant attenuation, yet both types of RF-based PASS devices that were tested had difficulty in reliably communicating with the base station. Radio interference, in particular from nearby high-power transmitters such as this, can present a potentially serious obstacle to RF-based PASS transmissions. Additional tests are being developed for measuring this impairment.

The data and corresponding discussion presented below are intended to aid in the development of laboratory-based test methods for RF-based emergency safety equipment such as RF-based PASS devices. Such test methods developed to date focus on inserting a controllable amount of attenuation between the portable PASS device and the PASS base station and inserting a specified level of RF interference between the portable and base station units. We also anticipate that additional test methods and standards will be forthcoming in the near future.

Analysis of Radio-Propagation Environments to Support Standards Development for RF-Based Electronic Safety Equipment

Kate A. Remley, William F. Young, Jacob Healy

Electromagnetics Division
National Institute of Standards and Technology
325 Broadway, Boulder, CO 80305

Abstract: *We analyze data from NIST field tests in which radio-propagation channel characteristics were measured at approximately the same physical locations as where the performance of various RF-based firefighter distress beacons was tested. These side-by-side tests were made in representative emergency responder environments, including an apartment building, four types of office buildings, a convention center, and an urban canyon. These environments contain propagation features that often impair radio communications, including stairwells, basements, and rooms deep within buildings, among others. The goal of this work is to determine appropriate performance metrics for use in the development of laboratory-based test methods for RF-based electronic safety equipment. For the structures we studied, we found that attenuation, rather than multipath, plays a more significant role in determining whether or not a remote distress alarm is received outside the structure. The analysis has enabled rough classification of structures into categories of attenuation values that can be used in laboratory-based test methods to verify the performance of the RF-based alarm system that we tested. The environments, tests, and measured data are discussed in detail.*

Key words: *attenuation; delay spread; emergency responders; firefighter communications; multipath; public-safety radio communications; radio propagation experiments; transfer function; urban canyon; wireless communications.*

1. Introduction

Emergency responders count on reliable radio communications between responders, who are often inside a structure, and the incident command station outside. New wireless technology is being developed that can further increase responders' safety and efficiency by remotely monitoring their position, status, and situational awareness. The responder community would like to take advantage of this technology. Because lives may depend on its performance, wireless technology used in emergency response scenarios must generally satisfy higher levels of reliability than technology used in the commercial sector. Even though standards currently exist for commercial wireless devices such as cell phones, wireless local-area-networks and handheld radios, at present few standards exist for wireless electronic safety equipment that primarily transmits data, as opposed to voice communications.

The U.S. Department of Homeland Security (DHS) Standards Branch has tasked researchers at the National Institute of Standards and Technology (NIST) with providing technical support for the development of consensus standards for these new products. As examples, DHS, through NIST, has determined gaps in existing standards and developed appropriate test methods for radio-frequency identification (RFID) systems used in public safety and government applications such as tracking or inventory control [1-3]. A second project is working with the urban search-and-rescue community to support development of standards for the wireless control of robots through ASTM International [4-5].

Here we describe DHS-sponsored work carried out to support the National Fire Protection Association (NFPA) in the revision of NFPA 1982: Standard on Personal Alert Safety Systems (PASS) [6] to include RF-based PASS. A PASS is essentially a "firefighter-down" alarm that emits a loud audible alarm when the wearer is motionless for 30 seconds. Some PASS manufacturers now include an RF transceiver in the body-worn PASS device to alert the incident command station. The transceiver is also capable of receiving an order from the incident command station to evacuate. The work presented here is expected to be applicable to other types of RF-based electronic safety equipment as it becomes available.

The technical strategy applied to this project is to first conduct field tests to gather information on typical values of key wireless-propagation-channel parameters in representative responder locations, including high-rise buildings, urban canyons, tunnels, apartment buildings, office buildings and other large structures where radio communication problems are sometimes encountered. Then, researchers determine representative values of these parameters and replicate the corresponding propagation-channel conditions in a laboratory-based, free-field test environment. The final step is to verify that the performance of a given wireless device in the laboratory is similar to that measured in the field. This process allows development of general, laboratory-based test methods that place the wireless device in conditions similar to those in which it will be used in the field. In this technical note, we describe field tests and analyze the measured results to develop appropriate performance metrics and their values for subsequent use in laboratory-based testing.

Laboratory-based test methods of RF-based emergency safety equipment provide the advantages of accuracy, repeatability, efficiency, and, often, reduced cost, when compared to the use of building structures and/or structure-based test beds. This is because we can carefully control the test environment and conditions in a laboratory while covering propagation-channel parameters measured over a wide range of building types. For the testing of RF-based equipment, we can expose the system under test to specific levels of attenuation, interference, or multipath, reproducibly and with known uncertainty.

However, reducing the complex, highly variable radio-propagation environment to a laboratory environment is extremely challenging. The approach taken here is typical of those reported in the literature [7-12]. Data were collected in several representative environments and then processed to extract the values of key parameters relevant to our specific end use.

5

It has been necessary for NIST to perform measurements as part of this project because much of the data that were previously published in the literature describe tests made to support commercial applications such as cellular telephone communications, where a base station provides coverage to a wide area, rather than the point-to-point, pedestrian-height scenarios utilized in many RF-based emergency scenarios. Most of the propagation-channel data analyzed here was collected in support of projects funded by NIST's Public Safety Communications Research Laboratory, in the NIST Office of Law Enforcement Standards [13-16 and the NIST Technical Notes referenced therein].

As mentioned above, one important goal of the study presented here was to analyze the performance of the RF-based PASS systems under the same conditions that the channel characterization tests were conducted. In the following sections, we discuss various aspects of our measurements that contribute to the uncertainty in the results of the measurement comparison. These aspects include: (a) the use of one frequency band for channel characterization, while the RF-based PASS systems operate in other bands, (b) the fact that the locations of the measurements were not identical, and (c) the use of reference measurements made in locations other than the field tests. Because of such nonidealities in our measurements, in some cases the performance of the RF-based PASS system disagrees with what is expected theoretically. However, certain trends are clearly indicated from the data, allowing us to identify representative values of attenuation and multipath for the development of laboratory-based test methods.

In Section 2, we describe the measurement system and data-processing algorithms that we used in the RF-propagation-channel characterization measurements. In Section 3, we describe the various environments in which the measurements were made. Section 4 contains the analysis, including a summary of the results of our measurements and a discussion of the relationship between RF-based PASS performance and propagation-channel characteristics. In Section 5, we discuss the assumptions and approximations that were made in measuring and analyzing the data and how they affect the uncertainty in relating RF-based PASS performance to propagation-channel characteristics. In Appendix A, we present a complete summary of the data that were collected in each environment, and in Appendix B, we present graphs that relate the success or failure of RF-based PASS transmissions to both attenuation and multipath.

2. Measurements of Path Loss and RMS Delay Spread

The goal of our analysis of the wireless-propagation-channel characterization measurements was to determine representative values of the key impairments to successful transmission of alarm signals from RF-based PASS systems. The two physical channel characteristics focused on in this study are attenuation (path loss) and multipath (reflectivity). Knowledge of the former is essential because the path loss, or reduction in signal strength experienced by a signal as it penetrates and travels through a structure, will directly impact the ability of an emergency responder to receive a signal from the incident command post, or vice versa. The level of multipath was also studied. In reflective environments, signals may travel from the transmitter to the receiver along multiple paths by way of signal bounces off metallic or other reflective surfaces. As a result, multiple copies of the signal may arrive at the receiver over a range of times. This effect may be described through a metric called root-mean-square (RMS) delay spread. For digitally modulated signals in particular, the self-interference arising from the multiple delayed copies of a signal can degrade the ability of the receiver to demodulate the received signal properly. Also, destructive interference may occur between signals following the various paths. The latter effect is termed fading.

To characterize both attenuation and multipath, we measured the wideband frequency response and time-delay characteristics of the outside-to-inside RF propagation channel with a measurement system based on a vector network analyzer (VNA), shown in Figure 2.1 below. This instrument collects data over a frequency range on the order of a communications channel (or wider) by stepping through frequencies one at a time. This system, described in more detail in [14], lets us measure the complex transfer function of the wireless-propagation channel as a function of frequency. In the field tests of [15, 16], data were typically acquired over a very wide frequency band (100 MHz to 18 GHz). We analyzed subsets of these data for the present report.

The procedure for the propagation-channel transfer function measurements is described in the following subsection. From the Fourier transform of the measured transfer function, the power delay profile (the received power as a function of time) and RMS delay spread of the channel were found in post processing. Our data cover several tens of megahertz, which is adequate to provide an estimate of the power delay profile over the bandwidths of interest. Because this range of frequencies is significantly wider than that of most modulated signals, we refer to these measurements as wideband.

The VNA acts as both transmitter and receiver in this system. The signal is amplified and fed to a transmit antenna, as shown in Figure 2.1. The signal propagates through the radio channel to the receive antenna. To characterize the one-way radio propagation channel, the received signal is returned to the VNA via a fiber-optic cable, where it is acquired and stored for post processing. The use of the fiber-optic cable eliminates the additional loss that would be introduced with a coaxial cable on the return path. One advantage of this system is that it provides a high dynamic range when compared to true time-domain-based measurement instruments. This is important because we typically measure weak signals in these experiments. One disadvantage is that a time-varying channel may change during the long acquisition period.

In Figure 2.1, the system is configured for a line-of-sight (LOS) reference measurement. In practice, the transmit and receive antennas may be separated by significant distances, although they must remain tethered together by the fiber-optic link. In the system we used, we could attain link distances up to 200 m.

We conducted two sets of measurements: One set over a "low"-frequency band that ranged from 100 MHz to 1.2 GHz; and one set over a "high"-frequency band, that ranged from 1 GHz to 18 GHz. The low-band measurements covered the operating frequencies of the PASS devices we tested, which operate in the 450 MHz and 900 MHz bands, and these data are used in the analysis that follows. The complete set of data may be found in [15].

For the propagation-channel characterization measurements, we used omnidirectional discone transmit and receive antennas. Omnidirectional antennas are often used with RF-based PASS base stations. The vertical beamwidth of the omnidirectional antennas is approximately 40 ° to 50 °. Identical antennas were used at transmit and receive sites.

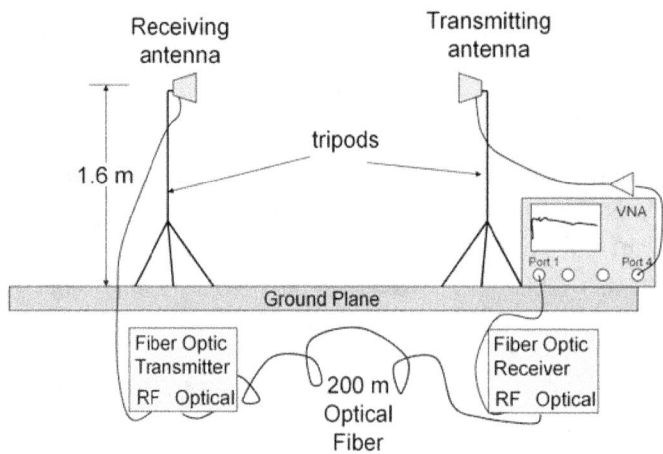

Figure 2.1: Wideband measurement system based on a vector network analyzer. Frequency-domain measurements of the complex RF propagation channel are facilitated by use of the optical fiber link. The measured data are transformed to the time domain in post-processing. Use of this system enables determination of path loss, time-delay spread, and other figures of merit important in characterizing modulated-signal transmissions.

To make a measurement, the VNA is first calibrated by use of standard techniques where known impedance standards are measured. The calibration enables correction for the response of the fiber-optic system, amplifiers, and any other passive elements and electronics used in the measurement. We also used a high-pass filter in post processing our measurements to suppress a large, low-frequency oscillation that occurs in the optical fiber link. Because the received signals measured during our field tests tend to be weak, an amplifier is used. Consequently, during calibration of the VNA, an attenuator is inserted in the "thru" calibration-standard path. This extra attenuation is corrected for in post processing the path-loss measurements. The value of the attenuator used is noted in each data set given in Appendix A.

For the measurements reported here, the VNA-based measurement system was set up with the following parameters: the initial output power was set to approximately −14 dBm. The gain of the amplifier and the optical link and the system losses resulted in a received power level at the VNA of no more than 0 dBm. An intermediate-frequency (IF) averaging bandwidth of around 1 kHz was used to average the received signal. The number of points varied for different measurements, but we generally aimed for a frequency spacing of 1 MHz to capture key propagation-channel response effects. The dwell time was approximately 20 μs per point. We next describe how the data acquired from the VNA measurements were processed to provide path loss and RMS delay spread.

2.1 Wideband frequency response and path loss

Our wideband measurements provide a complex channel transfer function $H(f)$, where $H(f)$ was given by the measured transmission parameter $S_{21}(f)$. Impedance mismatch was not considered, based on the assumption of well-matched antennas. To find the frequency-dependent path loss between the transmit and receive antennas, we first compute $|H(f)|^2/|H_r(f)|^2$, where $H_r(f)$ is a free-space reference transfer-function measurement made a known distance d_r from the transmit antenna. The use of a ratio to find the path loss enables us to correct for the transmit and receive antenna responses. Because these antenna responses are common to both the reference and the measurement, dividing one by the other removes the antenna effects from the measurement. We next correct the measurements for the free-space path loss between the transmit antenna and the reference location by dividing $|H_r(f)|^2$ by $(4\pi d_r/\lambda)^2$, where λ is the free-space wavelength at the frequency of interest. If this correction were not made, the measured path loss would be artificially reduced by the free-space path loss corresponding to distance d_r.

The reference transfer function may be acquired either during field tests or from a laboratory measurement. For the measurement data presented in Appendix A, laboratory reference measurements were used, and the reference measurement distance d_r is noted for each case.

Based on the above discussion, we calculate free-space path loss from our VNA measurements as (all quantities expressed in decibels):

$$\text{Path Loss} = 10*\log_{10}(|H(f)|^2/|H_r(f)|^2) + (\text{cal. attenuator value}) + 10*\log_{10}(4\pi d_r/\lambda)^2. \quad (2.1)$$

2.2 RMS delay spread

Root-mean-square (RMS) delay spread is calculated from the power-delay profile of a measured signal [17-19]. Figure 2.2 shows the power-delay profile for a typical building propagation measurement. The peak level usually occurs when the signal arrives at the receiving antenna, although sometimes the received signal builds up gradually to the peak value and then falls off (the latter behavior is indicative of a reverberant environment). Note that the dynamic range value and, consequently, the threshold value, may change for low levels of received signal. The following equation is used to define the RMS delay spread, σ_τ:

$$\sigma_\tau = \sqrt{\overline{\tau^2} - \left(\overline{\tau}\right)^2}. \quad (2.2)$$

In (2.2), $\overline{\tau}$ is defined as the average value of the power-delay profile in the defined dynamic range window and $\overline{\tau^2}$ is the variance of the power-delay profile within this window.

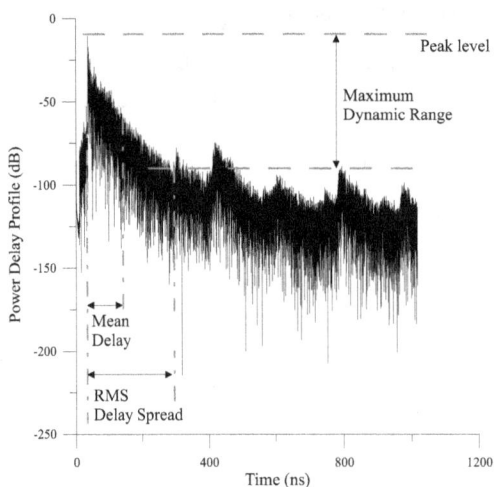

Figure 2.2: Power-delay profile for a building propagation measurement. Important parameters for a measured signal are the peak received signal power level, the maximum dynamic range (the difference in decibels between the peak and threshold values), the mean delay (the delay corresponding to the average value of received signal power above the chosen threshold), and the RMS delay spread (the second central moment of the received signal power above the threshold).

We find the RMS delay spread from the measured complex channel transfer function as follows. First, transfer functions were windowed with a Hamming window to reduce delay-domain sidelobes. This technique is often employed to process VNA measurements. Then, the windowed transfer functions were inverse-Fourier-transformed to obtain bandpass channel impulse responses. These bandpass channel impulse responses were then downconverted and low-pass filtered with a fifth-order elliptic filter to suppress the image frequencies. For a channel impulse response denoted $h(\tau, t_i)$, the corresponding i^{th} ("instantaneous") power delay profile (PDP) was computed as $P_i(\tau)=|h(\tau, t_i)|^2$, where τ denotes the decay time and t denotes the time at which the measurement was taken.

When multiple measurements were available from a particular site, we took the average of the instantaneous power delay profiles to compute the RMS delay spread. When only a single measurement was available, we found the "instantaneous" RMS delay, which provides a rough approximation of the RMS delay spread.

A common rule of thumb is to calculate the RMS delay spread from signals at least 10 dB above the noise floor of the measurement. The "noise floor" data in the graphs that follow were collected by terminating the transmitting port of the VNA in a 50 Ω load, so that the receiving port measured only background ambient signals. For the measurements described in the following sections, we used the method described in [16] to determine the useful dynamic range of each measurement. Where insufficient dynamic range existed, no RMS delay spread was calculated (represented by "N/A" in the tables of Appendix A).

Path loss (derived from the propagation-channel transfer function) and multipath (quantified by RMS delay spread), are two key propagation-channel characteristics that are predictive of the performance of wireless devices. These characteristics are analyzed in Section 4 for the various environments described in Section 3. A third propagation-channel characteristic, interference, will be tested and discussed in future work. Measurement uncertainty is discussed in Section 5.

3. Test Environments

We provide a brief overview of the environments and conditions in which radio-propagation-channel measurements were made and RF-based PASS systems were tested. The positions where the tests were conducted are marked on diagrams of each building, and are the same for both systems unless so noted. Photographs are provided to give the reader an indication of the characteristics of each environment. More information on these radio-propagation channel measurements can be found in NIST Technical Notes 1552 [15] and 1557 [16]. The environments are ordered with the environments presenting the lowest path loss first.

3.1 Denver urban canyon

Measurements were taken outdoors in the financial district of downtown Denver on two successive Saturdays in June, 2009. This area contains many buildings of over 20 stories. Figure 3.1(a) shows an illustration of the test area constructed from a Google map view.[1] Street widths were on the order of 20 m. For the VNA measurements, three transmitter (TX) locations and twelve receiver (RX) locations were tested, resulting in 36 sets of measurements [15]. RF PASS tests were carried out with the base stations located at the TX1 and TX2 sites. People and cars were moving through the test location during the measurements.

Results are presented here for six receive antenna locations for transmitter sites 1 and 2. Figure 3.1(b) shows a photograph of the VNA system low-band and high-band receive antennas located at position R5 on the corner of Welton and 17th Streets. The diagram in Figure 3.1(c) shows that the LOS distances ranged from 10 m to 80 m, with non-line-of-sight (NLOS) distances placed every 10 m past R5.

---

[1] © 2009 Google, Map Data © 2009 Tele Atlas.

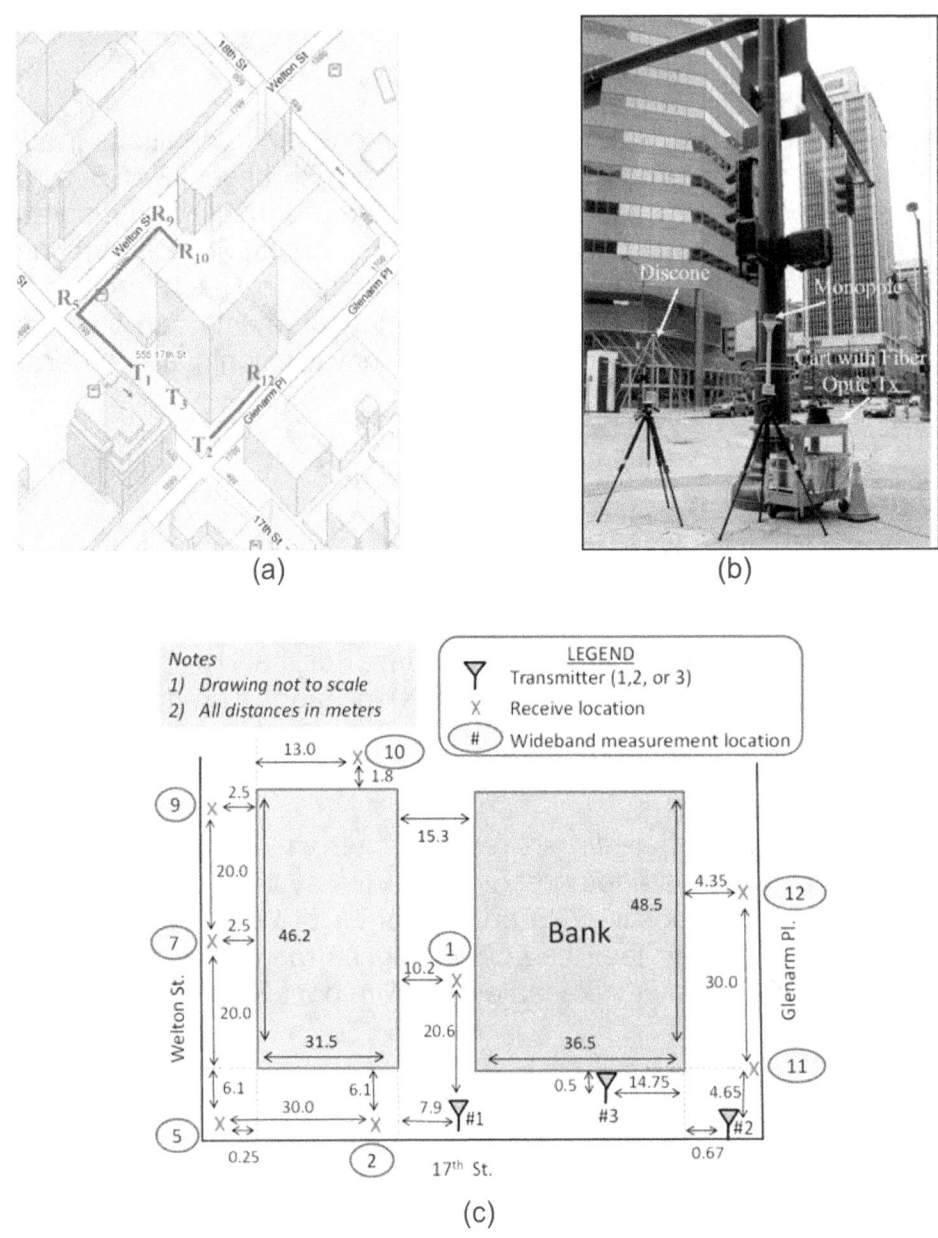

Figure 3.1: Denver urban canyon. (a) The path taken in the NIST measurements was down one block and around the corner. (b) Photograph of the wideband channel-characterization measurement system. (c) Locations of each measurement point.

3.2 Horizon West apartment building, Boulder, Colorado

The 12-story Horizon West apartment building in Boulder, Colorado is shown in Figure 3.2(a) and (b). The building is constructed of reinforced concrete, steel, and brick with standard interior finish materials. The building was fully furnished and occupied during the experiments. Measurements were performed during daytime hours and, as a result, people were moving throughout the building during the experiments.

The VNA transmit antenna and RF PASS base-station site was located approximately 60 m from the building, shown near the bottom of the diagrams in Figs. 3.2(c) and (d). The test positions are also shown in Figures 3.2(c) and (d). These measurements were acquired approximately every 5 m down the main hallways, as indicated in the figure, on Floors 2 and 7 of the building.

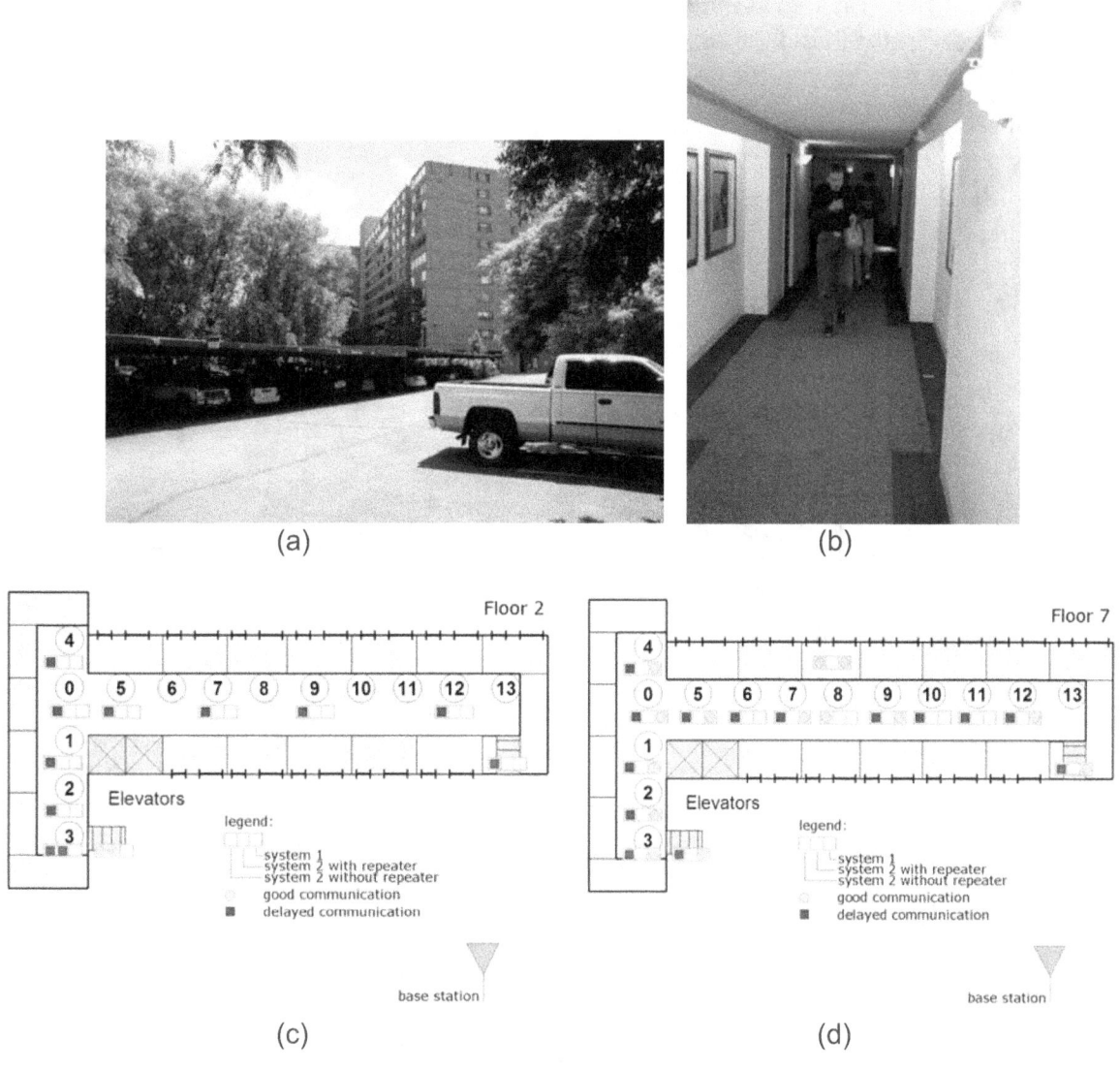

Figure 3.2: (a) 12-story apartment building Horizon West. (b) Inside on Floor 7. (c) Test positions on Floor 2. (d) Test positions on Floor 7. Blue squares indicate a delay in the reception of the alarm signal. Green circles indicate reception without significant delay.

3.3 Republic Plaza, Denver, Colorado

The Republic Plaza is a 57-story office building in downtown Denver, shown in Figure 3.3(a). The construction materials are a typical combination of concrete and steel. The exterior is a combination of glass and metal. The interior building materials consist of metal framing, drywall, and trim, with stone finishes in the lobby. The lobby is shown in Figure 3.3(b), and the 10th floor, which was in the process of renovation, is shown in Figure 3.3(c).

The VNA transmit site and RF PASS base station, depicted in the sketch of Figure 3.3(d), were located on the 17th Street side, approximately 10 m from the building. This location was intended to simulate the location of a command vehicle in an emergency response scenario.

Pink numbers on the sketch show the locations within the building where testing was conducted. The vertically stacked numbers indicate testing conducted in a stairwell. The highest floor tested was the tenth floor.

(a)

(b)

(c)

Figure 3.3: (a) 57-story Republic Plaza office building in downtown Denver, CO. (b) Main lobby and (c) tenth floor. (d) Test positions on the first 10 floors of the building. The pink numbers show the locations of the propagation-channel measurements and the green and red dots show the RF-based PASS performance.

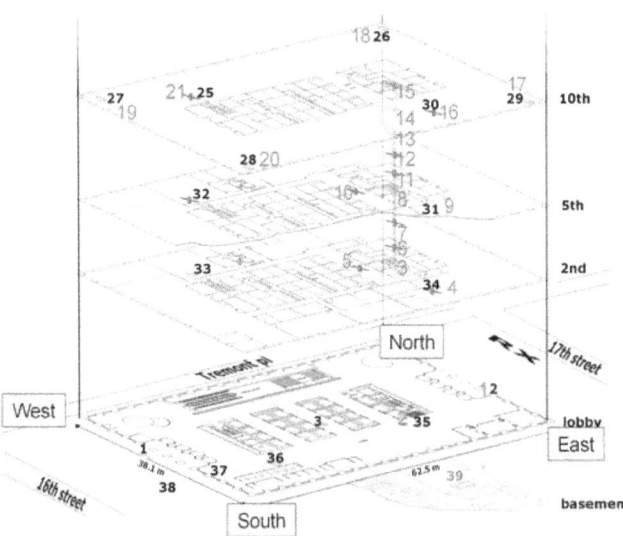

VNA measurement test locations are in pink

(d)

3.4 NIST Building 27

This structure consists of a small concrete building above ground connected by a subterranean tunnel to a small room. The front building consists of a room approximately 5.5 m (18.0 ft) wide and 7.1 m (23.3 ft) deep, shown in the photograph of Figure 3.4(a). There are two small windows in the main room. The room is used for storage and contains many boxes of electronics equipment, as shown in Figure 3.4(b). The room is connected to a much smaller room by a 24.5 m (80.4 ft) long tunnel, as shown in Figure 3.4(c). The tunnel and small room at the end, which is 3 m x 3 m (9.8 ft x 9.8 ft), are below ground and used to access the NIST open-area test site. The diagram in Figure 3.4(d) shows the dimensions of the building and the test positions.

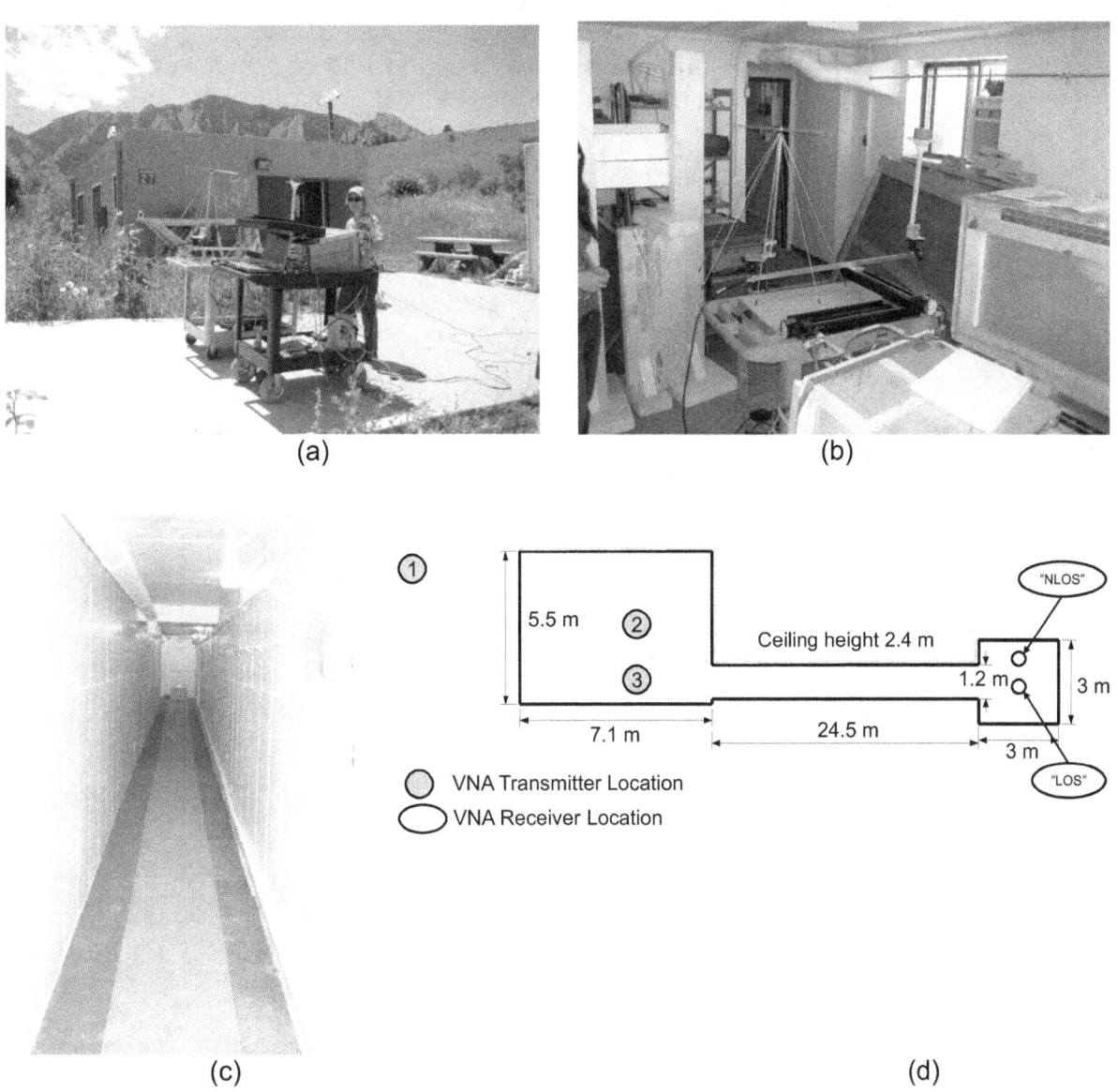

(a)   (b)

(c)   (d)

Figure 3.4: NIST Building 27. A small concrete main building ((a) outside and (b) inside) is connected through a tunnel (c) to a small room. The test positions are shown in (d).

3.5 NIST Building 24

This building consists of offices and laboratories, including a large semi-anechoic antenna test chamber approximately 25 m x 6 m. The building footprint is approximately 30 m x 30 m. The building is constructed of cinder block, concrete, and steel, as shown in Figure 3.5(a). There are few windows except in the offices and storage spaces. There are two levels above ground with offices and lab space. The building has a large, open, unfinished basement. This structure is similar to many small office buildings that may be encountered by emergency responders.

The VNA transmit antenna and RF PASS base station were set up immediately outside the building on the picnic table shown in Figure 3.5(b). Tests were conducted by entering the building at the door marked Position 1 in Figure 3.5(c), turning left and going down the stairs to the basement, walking to various sites throughout the basement (positions 2-7), including into an elevator at the end of a hallway, ascending the stairs (position 8), and walking down a corridor to the original entry position (positions 9 and 10).

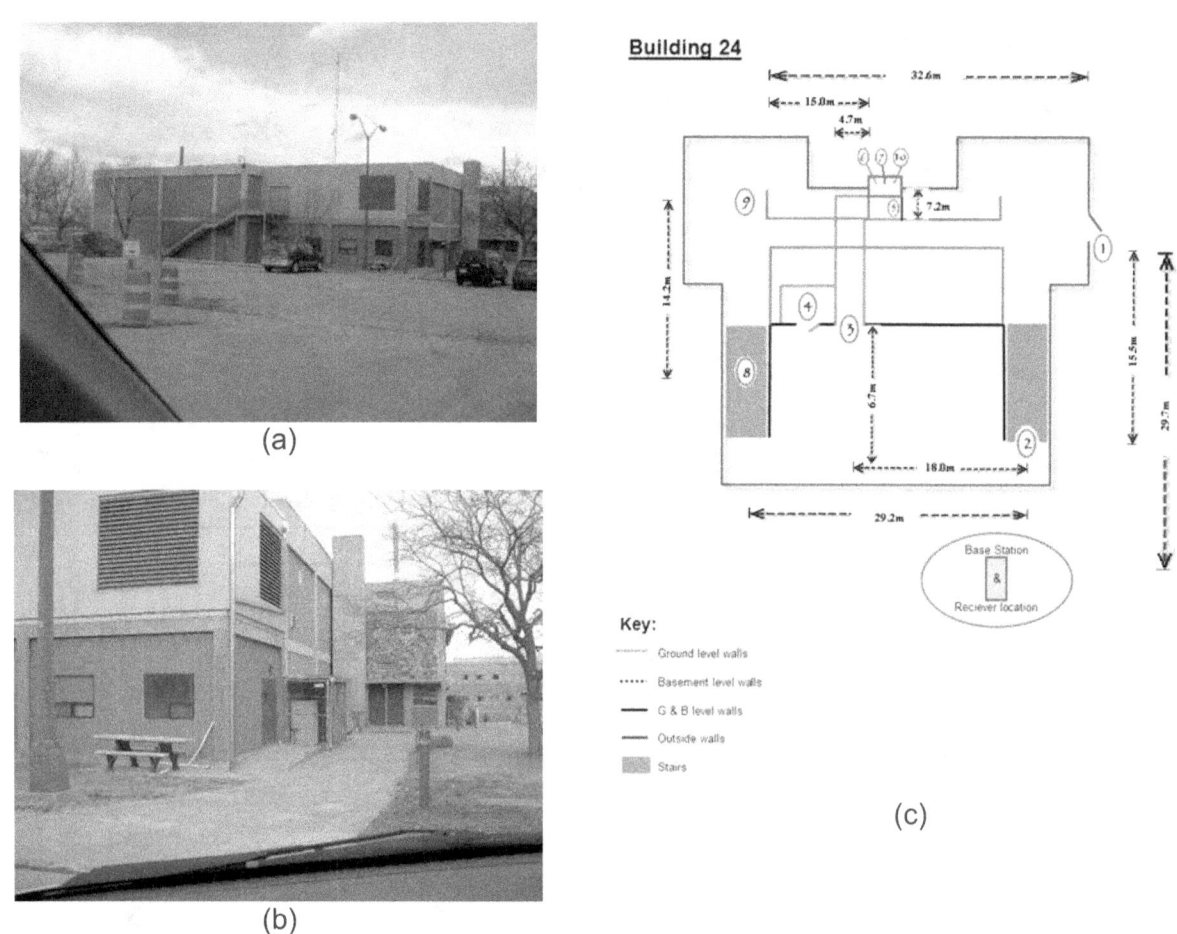

(a)

(b)

Building 24

Key:
- - - - - Ground level walls
· · · · · · Basement level walls
——— G & B level walls
——— Outside walls
▓▓▓ Stairs

(c)

Figure 3.5. NIST Building 24: (a) A two-story concrete building with a basement. (b) The VNA transmit antenna and RF-based PASS base station were set up on the picnic table. (c) Building layout and test positions on the first floor and basement.

3.6 NIST Building 1

This building is referred to as the Radio Building at the NIST laboratories in Boulder, CO. The building is constructed of reinforced concrete and is basically a four-story building consisting of six "wings" branching perpendicular to a main spine. Each wing consists of a corridor with a single lab or office on either side, as shown in Figure 3.6(a). The building is built on a hillside, and consequently, some locations in the building are below ground level. Measurements were made on the third-floor hallway called "Wing 4," continuing around the corner on the "main spine." The measurements were performed during the week in the daytime hours and, as a result, people were moving throughout the building during the experiments.

Two VNA transmit antenna/RF-based PASS base-station sites were assembled as shown in Figure 3.6(b). The site at Wing 4 was located on the loading dock, which is on the same level as Wing 4, while the site at Wing 6, shown in Figure 3.6(b), was approximately 10 m from the building and one level higher than Wing 4. Measurements were performed at the 18 locations indicated in Figure 3.6(c).

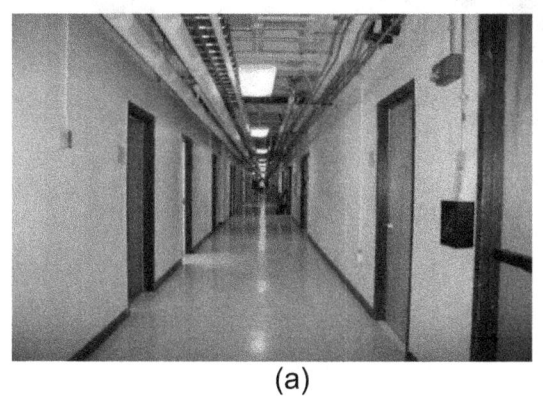

(a)

(b)

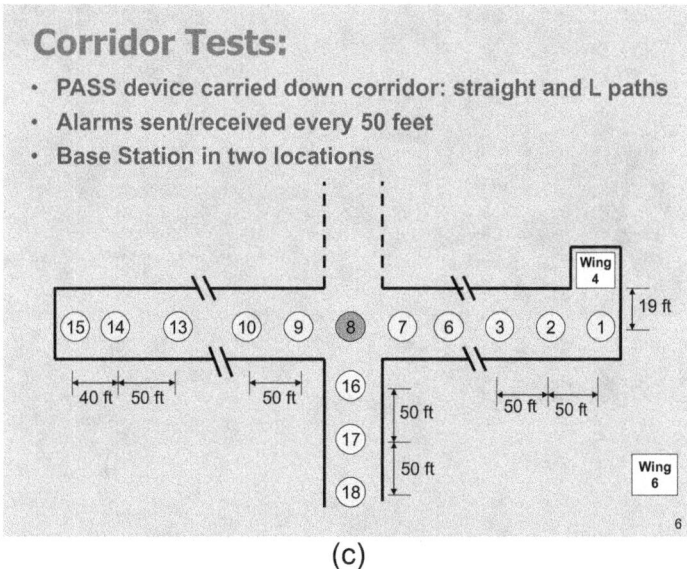

(c)

Figure 3.6: NIST Building 1. (a) Wing 4 hallway where test positions 1-7 were located. (b) Transmitter outside Wing 6, denoted by the yellow box in the diagram of (c). Tests were conducted with the RF-based PASS base station located at the end of Wing 4.

3.7 Colorado Convention Center

This massive three-level structure is constructed of reinforced concrete, steel, and standard interior finish materials, as shown in Figures 3.7(a) and (b). The exterior of the building is a combination of glass, metal, and concrete. Test positions are shown in Figures 3.7(c) and (d). As shown in Figure 3.7(d), the convention has a basement and two above-ground levels. Measurements were conducted when the convention center was empty of people.

The VNA transmit antenna and RF-based PASS base station were located approximately 10 m from the entrance on the Speer Boulevard side, shown by the yellow box labeled "RX" in Figure 3.6(c). PASS testing was conducted only at the positions marked by a letter in a green square. The channel measurements were made at the positions marked by a number in a yellow circle. Only one of the PASS systems was tested at the convention center location.

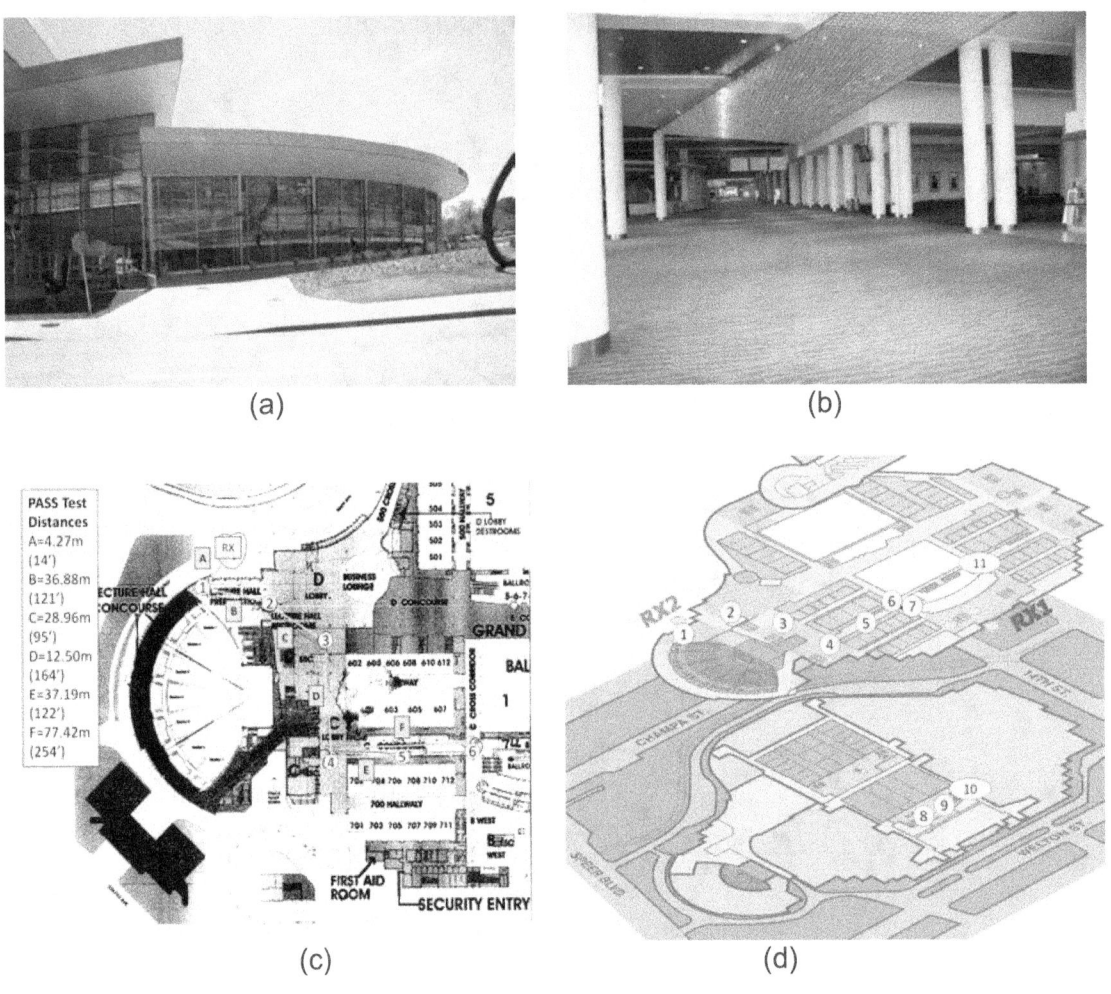

(a)

(b)

(c)

(d)

Figure 3.6: Colorado Convention Center. (a) Exterior, showing the location where the base station was positioned. (b) Interior, showing the large open spaces that were prevalent within the structure. (c) Top view of the test positions, including distances marked in the legend. (d) Three-dimensional view showing the main floor and basement.

4. Comparison of Channel Characteristics and RF-Based PASS Performance

In this section, we compare the results of the NIST measurements of path loss and RMS delay spread to the performance tests of the RF-based PASS made at approximately the same positions. This comparison is intended to allow us to identify representative values of propagation-channel characteristics for lab-based test methods of RF-based PASS and other RF-based electronic safety equipment. We extract from the measured VNA data the attenuation (path loss) and RMS delay spread (multipath) at the positions noted in the previous section for the various representative firefighter environments. Tables of these data are presented in Appendix A. In each table we note whether the RF-based PASS transmission was successful, experienced a delay (defined as a delay of more than one minute), or failed (defined as a delay greater than one minute).

In Section 4.1, we provide a brief summary of the data from the various environments discussed above. In Section 4.2, we study whether multipath or attenuation is the more critical impairment in a given environment for the successful transmission of an RF-based PASS alarm. To aid in this analysis we plot the RMS delay spread vs. path loss at each location within an environment in Appendix B. We indicate on each graph whether or not the RF-based PASS transmission was successful. Identifying which mechanisms have the most impact on PASS performance allows us to prioritize development of test methods.

In Section 4.3 we further analyze the data to classify propagation-channel environments in terms of measured levels of path loss and multipath. For each classification, we extract representative values of attenuation for laboratory-based RF-PASS testing.

The data presented below describe RF-based PASS performance with and without repeaters, at two different frequencies of operation, in various propagation-channel environments. The primary uncertainties in comparing our measured path loss data to device performance are assessed in the uncertainty analysis of Section 5. Even though the uncertainty is rather high, certain trends are clearly seen from the data, as discussed below.

4.1 Analysis of measured results

Here we summarize some of the key aspects of the measured data collected in various radio-propagation environments. As in Section 3, the order of the environments is based on lowest-to-highest path loss. Consult the tables in Appendix A and the graphs in Appendix B for more information.

We plot the RMS delay spread vs. path loss at each location within a structure in Appendix B. The success of an RF PASS transmission is indicated by a blue circle, the failure of a transmission (defined as a delay of three minutes or more) by a red x, and a significant delay (over one minute) by a green diamond.

- **Denver urban canyon – down street, around one corner**: This was the only outdoor-to-outdoor environment studied. Measured path loss was between 45 dB and 90 dB, but the RMS delay spread was as high as approximately 210 ns. The PASS devices operated successfully in this environment, except in two measurement locations, both having moderate path loss and RMS delay spread.

- **Horizon West – 12-story apartment building, Floors 2 and 7**: The path loss due to building penetration in this environment was not high, between 68 dB and 90 dB, and the RMS delay spread was less than ~55 ns on Floor 2 and ~80 ns on Floor 7. However, the RF-based PASS transmissions were generally not successfully received by the base station, with the exception of the 450 MHz system on Floor 7 locations, due to a cell-phone base station located on the roof of the apartment building. The strong cell-phone signal appears to have disrupted the RF-based PASS transmission, even when a repeater was deployed. The interference to both the portable unit and the repeater caused complete disruption of the transmission, as shown in Appendix A.2.

- **Republic Plaza – 57 story office building, Floors 1-10 including stairwell**: The plots of RMS delay spread vs. path loss in Appendix B clearly show that attenuation, not multipath, was the dominant channel impairment in this environment. When repeaters were used, reception was improved. The RMS delay spread values ranged from 50 ns to 400 ns.

  Note that the path loss values reported in the graph (~70 dB to 115 dB, confirmed by comparison to the single-frequency measurements shown in Figure 5.3 below and in [15]) are actually too small to cause the RF-based PASS to consistently fail. In the NIST lab, the RF-based PASS devices failed to successfully transmit an alarm at an attenuation level between approximately 100 dB and 135 dB, depending on the device. For the results shown in Figure B.3, we expect that the PASS base station antennas were not reoriented for maximum gain with respect to the remote units on the higher floors. The base station antennas have an elevation-dependent gain pattern. Consequently, the absolute values of path loss presented here have a higher uncertainty in the path-loss value at which the PASS communication fails. However, the results do clearly show that, even for the relatively high multipath in this structure (up to 400 ns RMS delay spread), attenuation was the primary cause of failure, as opposed to multipath.

- **NIST Building 27 – small main building connected by long subterranean tunnel to small back room**: Path-loss values ranged from around 85 dB to 100 dB, and RMS delay spread was generally low, but in one case it jumped to 250 ns, probably because of multipath in the front building before the signal propagated down the hall to the receiver. PASS measurements were not made in this structure.

- **NIST Building 24 – office/lab building with basement**: Measured path-loss values ranged from 95 dB to 115 dB. The attenuation in the basement was higher than 115 dB, but we were unable to measure it due to limited dynamic range of the VNA test set-up. In fact, the dynamic range was low enough that we were not able to use the outside-to-inside data to calculate the RMS delay spread except for one point. The RMS delay spread here was ~40 ns.

- **NIST Building 1 – office/lab building with long, part-subterranean corridor**: Measured path-loss values in this structure ranged from 100 dB to 140 dB and the RMS delay spread values were less than 100 ns. We were able to acquire meaningful data only at locations nearest the transmitter, because there was insufficient dynamic range to acquire path loss and RMS delay spread data. We expect that the path loss is significantly higher deeper inside the building. Where there was not enough dynamic range to calculate the RMS delay spread, we plotted it as zero. Because of the short RMS delay spread and large value of attenuation, we expect that attenuation is the primary failure mechanism in this environment for the RF-based PASS devices.

- **Colorado Convention Center – main entry, one corridor, and downstairs**: The measured path loss in this large structure was between 80 dB and 160 dB, and the RMS delay spread varied up to 180 ns. However, our measurement instruments had insufficient dynamic range far into the building, so we anticipate that the path loss was much higher than this. Path loss data were collected only to a lower frequency of 1 GHz. Because the path loss was calculated at this frequency and because a horn antenna (which is more directional than the RF-based PASS omnidirectional antenna) was used on the transmit side, relating RF-based PASS performance to values of path loss is difficult. Due to time constraints, we were able to measure only the 450 MHz RF-based PASS in this environment. RF-based PASS transmissions were received for a path loss of roughly ("roughly" for the reasons given above) 145 dB.

4.2 Channel impairments and RF-based PASS performance

For the environments we studied, the graphs in Appendix B clearly show that attenuation (path loss) is the dominant failure mechanism for the RF-based PASS system. In almost every case, there is a direct correlation between an increasing path loss and the failure of the RF-based PASS device transmission. Conversely, there seems to be little correlation between RMS delay spread and success or failure of the RF-based PASS. However, most of the environments we studied had relatively short values of 200 ns or less for RMS delay spread.

We conclude that, in the absence of external RF interference, lab-based tests that provide methods for testing RF-based PASS in a controlled attenuation environment will predict device performance in the majority of real-world firefighter environments. Tests utilizing various values of attenuation could be used to verify device performance in environments having the attenuation classifications listed in the table above.

Additional field tests and analysis should be conducted to determine the level of multipath in highly reflective environments such as factories, utility installations, and other manufacturing environments. Additional laboratory-based tests should be developed if it is found that these environments affect RF-based PASS performance.

Also, it is critical that interference tests be developed, because of the potential for RF interference to interrupt the RF-based PASS transmission, even when the size and composition of the environment should present no problem to successful reception.

4.3 Classifying levels of attenuation

Most of the environments we tested exhibited at least 50 dB of attenuation, created by the penetration of signals from outside-to-inside a structure (or vice versa), or the distance between transmit and receive antennas. Only the outdoor urban canyon environment and the shallow apartment building had maximum attenuation values less than 100 dB. We expect that typical house structures, small commercial buildings (such as small stores in strip malls and office buildings with exterior-facing offices) and small-to-moderate sized apartment buildings (in which all apartments have an exterior wall) would provide an environment where the total signal attenuation is less than 100 dB. We will classify this type of structure as "low attenuation," as shown in Table 4.1. With current (2012) technology, an individual RF-based PASS unit (no repeater) can operate successfully in these environments, unless external radio interference is experienced, as was the case in the Horizon West apartment building measurements.

Most of the environments we studied had maximum attenuation values between 100 dB and 150 dB, which we classify in Table 4.1 as "medium attenuation." We expect that the attenuation values in the Republic Plaza building and in the NIST Building 27 were on this order. We expect that moderate-sized structures such as small hospitals, and moderate-sized and tall commercial, office, and apartment buildings would provide an environment with attenuation between 100 dB and 150 dB. As can be seen in Appendix B, with current RF-based PASS technology, the use of a repeater can often overcome this level of attenuation.

Very large structures and those with subterranean floors, even of small size, can be expected to provide attenuation greater than 150 dB, which we classify as "high attenuation" in Table 4.1. NIST Buildings 24 and 1, and the convention center had such high levels of attenuation. We expect that multiple repeaters would need to be used in such environments, for current RF-based PASS technology. A summary of proposed path-loss classification is provided in the table below. As noted above, the RMS delay spread in the environments we studied did not exceed 200 ns. As a consequence, our classification focuses on attenuation rather than multipath. We expect that in a large factory environment, multipath may become a more significant problem.

Table 4.1: Classification of structures in terms of attenuation due to building signal penetration.

Classification	Attenuation (dB)	Typical structures	Current PASS
Low	Less than 100	Houses, small buildings with exterior-facing rooms	Single unit
Medium	100 to 150	Moderate-sized and tall structures with some interior rooms	With repeater
High	Over 150	Very large structures and those with subterranean floors	Multiple repeaters

5. Uncertainty in Relating Path Loss to RF-Based PASS Performance

Our analysis relates measured path loss and RMS delay spread to the performance of various RF-based PASS systems. In this section, we provide an estimate of the uncertainties in our path-loss measurements, combined with the additional systematic error arising from the use of path-loss measurements made in the 750 MHz frequency band (ranging from 725 MHz to 800 MHz), when the RF-based PASS systems operate in the 450 MHz (ranging from 400 to 500 MHz) or 900 MHz (ranging from 902 MHz to 928 MHz) frequency bands. We focus on the uncertainty in path loss because our analysis indicated it was the primary cause of failure in RF-based PASS transmissions.

Following the convention described in [20], the uncertainties associated with the measurement and estimation of path loss can be broken into two categories: Type A (evaluated by statistical means) and Type B (evaluated by non-statistical means). Contributions associated with time and location variation in the channel and the repeatability of the measurement instrumentation are described with Type A techniques. Systematic effects are described with Type B methods. These include errors in estimates of path loss at 450 MHz and 900 MHz with measured path-loss data acquired at 750 MHz, the use of reference measurements made in a controlled environment rather than at the field site, and drift of the measurement instrumentation. We describe these effects below, and then calculate the combined expected uncertainty in our estimation of path loss due to these contributions.

5.1 Small-scale fading

A key source of random, Type A, uncertainty in our estimate of the path loss can be attributed to small-scale fading, often called channel variability in the literature. Small-scale fading occurs from multiple frequency-, time-, and position-dependent reflections in the local area around each test location. Even though a building environment is fixed and measurements made there would be deterministic, small-scale fading is considered random due to its extreme sensitivity to antenna placement and the fact that cars, trucks, and pedestrians move randomly through the environment during measurement.

Small-scale fading manifests itself as rapidly varying noise-like peaks and nulls in the received signal strength. This type of fading is superimposed upon a "large-scale" mean value, where the mean is typically calculated over frequency, time, and/or a local position of one or two wavelengths. An example of small-scale fading as a function of frequency can be seen in Figure 5.1(a) and (b), where the top curves show the propagation-channel transfer function and the bottom curves show the noise floor of the measurement (described on p. 14). Increasing the number of measurements over a local area, over time, or over an increasing number of frequencies will typically reduce the uncertainty due to small-scale fading on an estimate of the path loss.

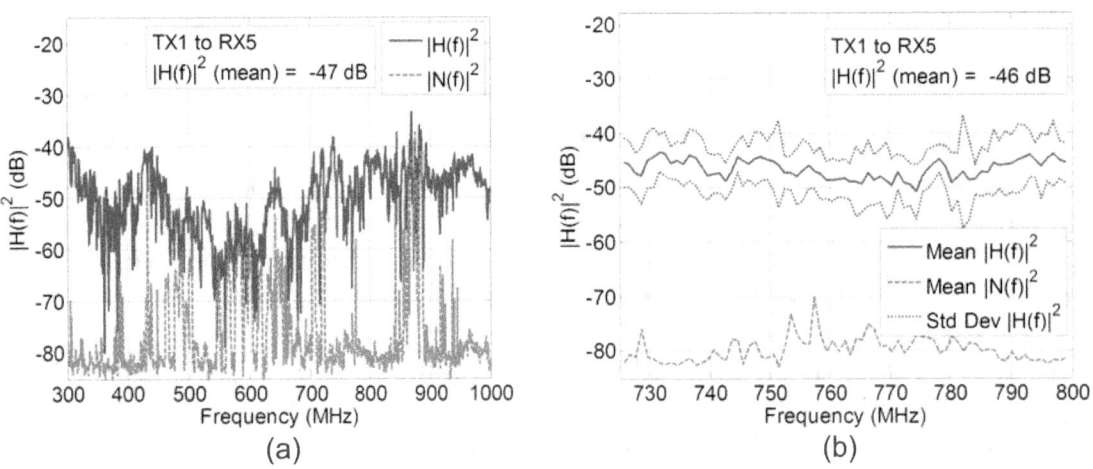

Figure 5.1: Propagation-channel transfer function measurements measurements made in the Denver urban canyon [16]. The lower red curves in the graphs show the "noise floor." Graph (a) shows the frequency range from 300 MHz to 1 GHz. The large spikes correspond to interference from radio signals present in the environment. Interferers can be seen in the mid-400 MHz band and the 900 MHz band. Graph (b) shows a representative measurement (one of 36) in the 725 MHz to 800 MHz band used for the path loss calculations. Nine measurements were taken at each transmit/receive antenna location in the 750 MHz band by moving the tranmit antenna on a 0.5 m x 0.5 m grid. These measurements were repeated once. The mean and standard deviation of the 18 measurements are indicated in the figure.

Measurements made in the Denver urban canyon (described in more detail in Section 3 and [16]) allow us to estimate the uncertainty in our estimate of path loss due to small-scale fading. For these measurements, the mean path loss at each of 12 receive antenna locations was estimated from 18 measurements (two measurements at each of nine antenna-positioner locations per receive site). An example of data from one receive site is shown in Figure 5.1(b), where we have plotted the mean of 18 measurements as a function of frequency. The standard deviation of these 18 measurements is also shown.

To estimate the uncertainty in the path loss, we first found the standard deviation at each of the nine antenna positioner locations over the frequency band 725 to 800 MHz (75 frequencies in 1 MHz frequency increments). We then calculated the standard deviation of these values over all nine antenna-positioner locations, including the second, repeat measurement. These two standard deviations, labeled "spacing" and "freq" are provided in Table 2.2. We see that the "combined" standard deviation for each receive antenna location in Table 2.2 is generally between 5 dB and 6 dB. When we combine all 36 values (12 receive antenna locations x three transmit antenna locations), we obtain an uncertainty u_{fading} in an estimate of path loss made at a single location due to small-scale fading of 5.7 dB.

Table 2.2: Standard deviation for 18 path-loss measurements at each of 12 locations. The "combined" column is the root-sum-of-squares combination of the linear values corresponding to the "spacing" and "frequency" standard deviations.

RX Location	725-800 MHz – all values in dB								
	TX1 σ			TX2 σ			TX3 σ		
	freq	spacing	combined	freq	spacing	combined	freq	spacing	combined
1	5.85	1.41	6.11	5.61	0.89	5.84	5.79	0.81	6.00
2	3.41	0.78	3.98	3.35	1.18	4.03	4.22	1.77	4.83
3	5.61	1.21	5.88	5.27	1.09	5.57	4.26	1.45	4.79
4	5.17	1.40	5.52	4.88	1.16	5.24	5.04	1.44	5.42
5	5.73	1.03	5.97	4.97	1.13	5.31	4.77	1.31	5.17
6	5.72	0.89	5.94	6.27	0.79	6.44	6.13	1.10	6.33
7	5.72	0.91	5.94	5.69	1.02	5.93	5.32	1.05	5.60
8	5.65	0.81	5.87	5.79	0.74	6.00	5.52	0.92	5.77
9	5.37	0.70	5.61	6.18	1.71	6.44	6.04	1.02	6.25
10	6.01	0.95	6.21	5.48	1.10	5.75	5.97	0.87	6.00
11	5.73	1.17	5.98	2.41	1.27	3.42	5.78	1.40	4.83
12	5.21	1.03	5.51	5.84	1.11	6.07	5.98	0.88	4.79

5.2 Repeatability of measurement instrumentation

To quantify the repeatability of the VNA measurements, which is also a Type A uncertainty, we conducted multiple measurements at the NIST Open Area Test Site (OATS). This is a 30 m x 60 m ground plane located many electrical wavelengths from the nearest reflective objects or scatterers. We performed a set of reference measurements, that is, direct line-of-sight measurements between the transmit and receive antennas. We used the same antennas and measurement set-up used in the Denver urban canyon environment, with the exception of a longer coaxial cable between the antenna and the VNA. Measurements were collected at 2 m increments for antenna separations between 4 m and 10 m in the 750 MHz band. The antennas were located 5 m above the ground, resulting in little contribution from ground reflections. The separation between antennas was measured with a tape measure, which likely increased the standard deviation reported below. Two sets of measurements were performed covering the 700 MHz band. One set of data covered frequencies from

725 MHz to 800 MHz. The second set of measurements was conducted between 300 MHz and 1 GHz. From this wideband measurement, data around the 750 MHz band were extracted.

We calculated the standard deviation in the path loss for this series of reference measurements by subtracting off the expected free-space path loss for each antenna separation value relative to the 2 m case, and then finding the standard deviation in the remaining path loss values. We then computed the standard deviation of the eight measurements (four antenna separations times two measurements). The standard deviation for the eight measurements was approximately 0.3 dB. Thus, the estimated Type A uncertainty u_{repeat} for the VNA measurement system is given as 0.3 dB for the 750 MHz band.

5.3 Off-frequency estimates of path loss

As discussed above, the path loss and RMS delay spread data presented in Appendices A and B were calculated from measured data collected over the frequency band 725 MHz to 800 MHz (unless otherwise noted), rather than at the operating frequencies of the RF-based PASS devices. We used these "750 MHz band" data for various reasons: First, the 450 MHz and 900 MHz bands are heavily utilized by other wireless equipment, making it difficult to collect propagation-channel data that are unaffected by external radio interference, as shown in Figure 5.1(a). Also, NIST has collected a significant amount of propagation-channel data in the 750 MHz frequency band as part of a study of emergency responder use of this newly allocated frequency band [15, 16]. We decided to utilize the data collected for that project in our study of RF-based PASS performance, simplifying the data collection and analysis procedure significantly. Use of these off-frequency path loss data will introduce a systematic error when we relate the RF-based PASS device performance (success, failure, or delay) to a path loss value measured at the same location. We quantify this additional source of uncertainty by comparing measured path loss for various environments at the frequencies of interest (450 MHz, 750 MHz, and 900 MHz).

We first consider theoretical free-space path loss at the three frequencies. The ideal free-space path loss value is plotted in Figure 5.2, where path loss is given by

$$PL = -10\log_{10}\frac{\lambda^2}{(4\pi d)^2}.$$
(5.1)

As shown in Figure 5.2, when compared to a path loss measurement at 750 MHz, we would expect approximately 4 dB less path loss at 460 MHz and 2 dB more path loss at 900 MHz.

We next consider path loss determined from NIST measurements made in two different, non-free-space environments. Figures 5.3(a) and 5.3(b) provide an estimate of the attenuation versus frequency behavior for two representative scenarios: (a) a 60-story high-rise office building in Denver, CO [15], and (b) the outdoor urban canyon setting of Denver, CO discussed above and in [16]. In both cases, empirical cumulative distribution functions (CDF) are estimated from the received signal power that was measured at multiple fixed sites.

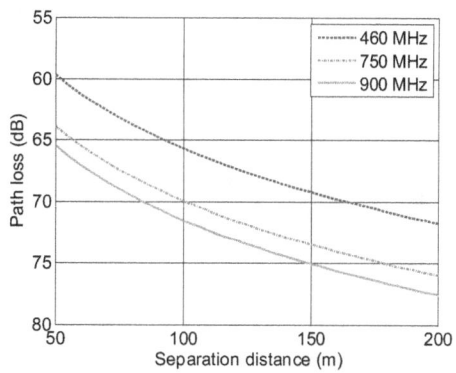

Figure 5.2: Simulated free-space path loss, from equation (5.1), as a function of transmit and receive antenna separation at three frequencies. At a separation of 200 m, the path loss values are 71.2 dB at 460 MHz, 76.0 dB at 750 MHz, and 77.5 dB at 900 MHz.

Collection and processing for these data were performed as follows: continuous-wave (CW) sources (handheld radios transmitting an unmodulated carrier) were carried through the building or urban streets, and a spectrum analyzer connected to an antenna measured the received power for a single frequency. Use of narrowband filters and a check of background environmental signals ensured that the signals we acquired corresponded to the ones we transmitted, rather than those from interfering sources.

Different handheld radio transmitters were used for each of the various frequencies. Measurements from all radios were corrected for nominal differences in transmit power levels (1 W or 5 W, depending on the frequency). This correction provides us with an order-of-magnitude estimate of the difference in path loss from one frequency to another for these environments. More detail on these measurements can be found in [15] and [16].

The corrected data acquired at all of the various receive sites were combined for each frequency, and a Kaplan-Meier estimate of the CDF was calculated for the data set. This estimated CDF was then used to obtain the parameters for a fit to a log-normal CDF. The log-normal distribution was used for the fit because it represents the typical power distribution in cluttered environments. The CDFs for the various frequencies are shown in Figures 5.3(a) and 5.3(b), with the received power levels for threshold values of 0.5 listed in the figures. For the frequencies of interest here, the 430 MHz, 750 MHz, 900 MHz, and 2.4 GHz bands, the root-mean-square error (RMSE) between the collected data and the estimated CDF curves is less than 6%.

Power levels at the 0.5 CDF threshold values show the frequency dependence of attenuation in these two representative environments. For example, for the 60-story office building (Figure 5.3(a)), the 2.4 GHz signal was received with a signal strength greater than −99.7 dBm only 50 % of the time, whereas the 750 MHz signal was greater than −86.9 dBm at least 50 % of the time. Thus, for this case, the 750 MHz signal was attenuated approximately 12.8 dB less than the 2.4 GHz signal.

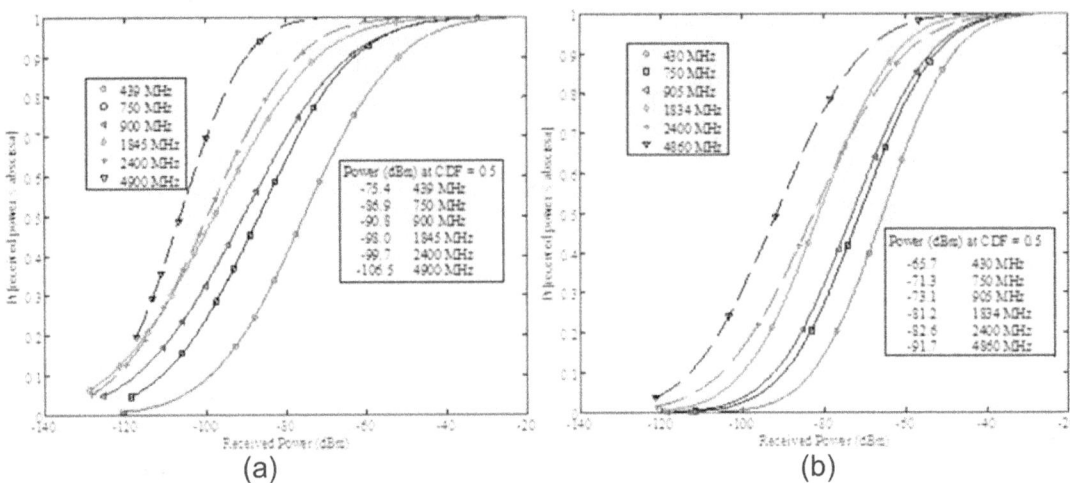

(a)   (b)

Figure 5.3: Cumulative distribution functions for various single-frequency measurements made at a large number of locations throughout two different radio-propagation environments: (a) a 60-story high-rise office building and (b) the Denver urban canyon. The received power at a threshold of 0.5 is shown in the legend, providing an estimate of the expected differences in path loss for different RF-based PASS carrier frequencies.

Table 5.2 shows the 0.5 CDF threshold values for frequencies of interest for the RF-based PASS systems we tested, along with data from 2.4 GHz, which is the frequency of operation for many wireless devices. The 750 MHz path loss value is used as a reference, and the various rows show the difference between the reference path loss and path loss values at the frequencies of interest. For example, when the transmit and receive antennas are placed at the same locations in the high-rise office building, the 450 MHz signal would be expected to experience 10.9 dB less path loss on average than the 750 MHz signal, while the 900 MHz signal would experience 1.1 dB more path loss on average. Note that a wireless system operating in the 2.4 GHz frequency band would need to overcome approximately 25.4 dB more path loss than one operating at 450 MHz in the high-rise office building environment, and approximately 11.3 dB in the urban canyon.

Table 5.2: Comparison of CDF values averaged over a 90 % interval (that is, for probabilities from 0.05 to 0.95) The "Difference" column provides an estimate of the expected difference in path loss for a given frequency band as compared to the reference frequency band of 750 MHz.

Frequency Band	Free Space	High-Rise Office Building		Urban Canyon		NIST Laboratory Building	
	Difference (dB)	Average Power (dBm)	Difference (dB)	Average Power (dBm)	Difference (dB)	Average Power (dBm)	Difference (dB)
750 MHz (reference)	0	-70.0	0	-59.2	0	-74.8	0
450 MHz	+4.8	-59.1	+10.9	-54.8	+4.4	-75.8	-1.0
900 MHz	-1.5	-71.1	-1.1	-61.0	-1.8	-76.6	-1.8
2.4 GHz	-10.1	-84.5	-14.5	-66.1	-6.9	-88.3	-13.5

Note that the high-rise office building represents a worse-than-average case because, as the transmitters were carried to higher floors in the building, the frequency-dependent receive antenna gain pattern had more of an impact on the received signal level. We were, in this case, measuring a combination of the path loss and the antenna pattern, which was not our goal.

Taking the median value of the four path loss differences corresponding to free space, the high-rise office building, the urban canyon, and the laboratory building, we expect to overestimate the path loss in the 450 MHz band by around 4.6 dB, and we expect to underestimate the path loss values in the 900 MHz band by around 1.6 dB. (We use the median rather than the mean to avoid dominating the estimate by the extreme values, and because we have only four samples.) For convenience, we prefer to use one value to approximate the error in our estimate of RF-based PASS performance at either frequency. Thus, we use the larger of the two values, giving ±4.6 dB for u_{path}. This represents the uncertainty introduced into an estimate of RF-based PASS performance at 450 MHz or 900 MHz by the use of a 750 MHz band path-loss measurement. This is a systematic Type B contribution to the uncertainty.

5.4 Controlled-environment reference measurement

As described in Section 2.1, to calibrate out the frequency-dependent loss (or gain) of the antenna and connecting cables from out path-loss measurements, we conducted a line-of-sight reference measurement. Such a reference measurement is made by placing the receive antenna a known distance from the transmit antenna with the entire measurement system assembled (including the optical fiber, the amplifier, and all cables to be used in the field tests) and conducting a VNA measurement. Subsequent measurements made at the field test location are corrected with the reference measurement by use of equation (2.1).

Reference measurements made at a field test location can include the effects of environmental reflections and interferers. As a result, we wish to conduct a reference measurement in a controlled environment. To assess the impact of the environment on our reference measurements, we compared reference measurements made in the heavily cluttered and dynamically changing Denver urban canyon environment, shown in Figure 5.4(a) and described in the following sections, to those made at the NIST open-area test site (OATS), shown in Figure 5.4(b).

(a)   (b)

Figure 5.4: Reference measurements made in two environments. (a) Denver urban canyon, with high-frequency-band antennas placed approximately 4 m apart and 1.5 m high. (b) NIST open-area test site, with low-frequency-band antennas placed 3 m apart and 5 m high.

Figure 5.5 shows two reference measurements covering the frequency range from 100 MHz to 1.2 GHz. The solid line shows the 4 m reference made in the Denver urban canyon and the dotted line shows a 4 m reference made in the controlled environment of the NIST open-area test site. The controlled reference measurement is smoother and does not show the dips around 400 MHz, 600 MHz, and 1 GHz that the reference made in downtown Denver shows.

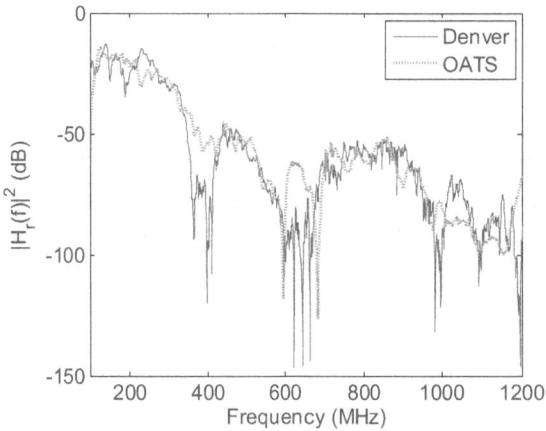

Figure 5.5: Graphs showing two different 4 m reference files of the type that would be used to determine $|H(f)|^2$ over the frequency range 100 MHz to 1.2 GHz. The measurements were made in (a) Denver, CO urban canyon, and (b) NIST open-area test site.

Note that the transfer function $|H(f)|^2$ would not be significantly affected by the reference except for cases where the received signal was at a level comparable to that of as the reference; that is, for cases with low signal-to-noise ratio. We used reference measurements made in the controlled environment of the OATS wherever possible. To utilize the controlled-environment reference measurement for all measurements reported in the following sections was not possible. As a result, in Appendix A, the type of reference measurement is noted for each environment.

Table 5.3 illustrates the difference in estimated path loss when two different reference files were used. These data correspond to measurements at five locations on the second floor of the Horizon West apartment building, Boulder, CO. This comparison was done in the 725 MHz to 800 MHz frequency band. The differences in the estimated path loss values are less than 0.1 dB. Thus, we neglect this source of uncertainty u_{ref} in our estimate of path loss.

Table 5.3: Comparison of path-loss estimates from measurements made at five different receiver locations in a 13-story apartment building when different reference files are used.

Receiver Location	Field Reference (dB)	Controlled-Environment Reference (dB)
1	86.13	86.13
2	82.78	82.79
3	77.66	77.67
4	78.34	78.35
5	91.12	91.12

5.5 Systematic measurement-system uncertainties

We next consider the Type B uncertainties of the VNA measurement system. The Type B uncertainties include the impact of temperature changes on measurements of S_{21}. The analysis in [21] estimated a VNA drift over three days in a laboratory setting as 0.1 dB, or u_{drift} = 0.1 dB. S_{21} data collected on the fiber optic link over several days in a laboratory setting provides an estimated Type B uncertainty associated with the fiber link at 0.1 dB, that is, u_{fiber} = 0.1 dB [16].

5.6 Combined uncertainty in relating path-loss to RF-based PASS performance

Table 5.4 summarizes the various components of uncertainty described in the previous subsections. The combined uncertainty for our estimates of path loss used to assess wireless devices operating at 450 MHz and 900 MHz is

$$u_{combined} = \sqrt{u_{fading}^2 + u_{repeat}^2 + u_{750MHz}^2 + u_{ref}^2 + u_{drift}^2 + u_{fiber}^2} \quad . \tag{5.2}$$

These uncertainties were combined by use of a root-sum-of-squares on the linear (as opposed to logarithmic) values, and then converted back to decibels, giving

u_combined of 7.0 dB. This uncertainty is high, but the goal of this work is to provide broad classifications of environments whose measured path-loss values range from a few tens of decibels to well over 100 decibels. Thus, an uncertainty of 7.0 dB is acceptable for our application. The errors in our estimate of RMS delay spread are likely significant as well. However, because the data indicate that path loss, not multipath, is the most significant impediment to RF-based PASS performance, we did not consider these errors in the present work.

Table 5.4: Uncertainties in estimated path loss values that are used to assess RF-based PASS performance at 450 MHz and 900 MHz.

Uncertainty Name and Type	Uncertainty Description	Method of Estimate	Value (dB)
u_fading Type A	Small-scale fading ("channel variability") derived from multiple path loss estimates.	Mean of the standard deviation from 36 independent path loss estimates. Each standard deviation found from 75 frequencies, nine antenna positions, and two repeats.	5.7
u_repeat Type A	Measurement system repeatability.	Standard deviation from eight independent reference measurements at NIST OATS. Derived from complete system measurements, including VNA, fiber optic link, and antennas.	0.3
u_750MHz Type B	Use of 750 MHz path loss measurements to estimate path loss at 450 MHz and 900 MHz.	Median difference between the average power over the 90 % probability intervals at 750 MHz and other frequencies in three representative environments and free space.	4.6
u_ref Type B	Use of reference measurement made in controlled environment on field test data.	Difference in path loss estimate for representative environment calculated with both types of reference data.	neglected
u_drift Type B	Drift in VNA measurements over time	Observed VNA drift over three days.	0.1
u_fiber Type B	Impact of temperature on fiber optic cable	Observation in controlled experiment over three days.	0.1

6. Summary and Conclusion

We compared measurements of the path loss and RMS delay spread of radio-propagation channels to the performance of RF-based PASS systems in a number of environments. The tests were conducted in building structures representative of those in which firefighters and other emergency responders are expected to have difficulty with radio communication. RF-based PASS performance was tested at approximately the same locations that the propagation-channel characterization was carried out. The base station and portable RF-based PASS unit were deployed to mimic the building penetration scenario of the firefighter-to-incident command transmission. We analyzed the measured data to determine the primary impairments to successful transmission of alarms, and to extract representative values of these impairments for use in laboratory-based test methods.

The majority of data that we analyzed indicate that signal attenuation, rather than multipath, is the dominant impairment to RF-based PASS transmission when no external RF interference is present. Our analysis of the data also allowed us to develop a rough classification of various building types in terms of attenuation. Laboratory-based test methods utilizing the values of attenuation identified in this study are currently being developed to verify device performance of RF-based PASS in various environments. These test methods are intended to be used in consensus standards to ensure that RF-based PASS systems operate reliably in the presence of a certain amount of path loss.

Note that, with current technology, RF interference can present an insurmountable obstacle to successful transmission of alarm signals, no matter what the attenuation classification of the structure. Laboratory-based tests are also being developed that assess device performance in a controlled environment under RF interference conditions.

Additional field tests and analysis should be conducted to determine the level of multipath in highly reflective environments such as factories, utility installations, and other manufacturing environments. Laboratory-based tests should be developed if these environments are found to impact RF-based PASS performance.

The authors hope that the data presented here, along with future sets of data, can be used to develop a complete suite of test methods, not only for RF-based PASS systems, but also for other RF-based electronic safety equipment. The impairments to RF-based wireless devices described here are general and could be used to develop standards for other equipment as the need for standards arises for these systems.

Acknowledgements

The analysis and development of test methods for RF-based emergency equipment has been funded under interagency agreement HSHQDC-11-X-00497 by the Department of Homeland Security, Science and Technology Directorate, Standards Branch, Philip Mattson, Chief and Bert Coursey (former Chief). The RF propagation-channel studies were funded by the Public Safety Communications Research (PSCR) program, a joint program of the National Institute of Standards and Technology (NIST) Office of Law Enforcement Standards (OLES) and the National Telecommunications and Information Administration (NTIA) Institute for Telecommunication Sciences (ITS). Dereck Orr, Program Manager. Jeff Bratcher, Technical Manager. Jason Coder, Galen Koepke, Chris Holloway, John Ladbury, Dennis Camell, and Camillo Gentile of NIST and Chriss Grosvenor of NTIA participated in the measurements. David Matolak, Qian Zhang, and Qiong Wu of Ohio University participated in the measurements in the Denver urban canyon and on the NIST campus.

References

[1] D.R. Novotny, J.R. Guerrieri, and D.G. Kuester, "Potential interference issues between FCC Part 15 compliant UHF ISM emitters and equipment passing standard immunity testing requirements," *IEEE EMC Symp. Dig.*, Aug. 2009, pp. 161-165.

[2] M.R. Souryal, D.R. Novotny, J.R. Guerrieri, D.G. Kuester, and K.A. Remley, "Impact of RF interference between a passive RFID system and a frequency hopping communications system in the 900 MHz ISM band," *IEEE EMC Symp.* Dig., July 2010, pp. 495-500.

[3] K.A. Remley, M.R. Souryal, W.F. Young, D.G. Kuester, D.R. Novotny, and J.R. Guerrieri, "Interference tests for 900 MHz frequency-hopping public-safety wireless devices," *IEEE EMC Symp. Dig.*, Aug. 2011, pp. 497-502.

[4] National Institute of Standards and Technology, "Statement of Requirements for Urban Search and Rescue Robot Performance Standards-Preliminary Report"http://www.isd.mel.nist.gov/US&R_Robot_Standards/Requirements%20Report%20(prelim).pdf .

[5] K.A. Remley, G. Koepke, E. Messina, A. Jacoff, and G. Hough, "Standards development for wireless Communications for Urban Search and Rescue robots," *Proc. Intl. Symp. Advanced Radio Technologies*, June 2007, pp. 66-70.

[6] National Fire Protection Association, NFPA 1982: Standard on Personal Alert Safety Systems (PASS), document scope available at http://www.nfpa.org/aboutthecodes/AboutTheCodes.asp?DocNum=1982 , accessed Nov. 4, 2011.

[7] A. A. M. Saleh, R. A. Valenzuela, "A Statistical Model for Indoor Multipath Propagation," *IEEE Journ. Selected Areas Comm.*, vol. SAC-5, no. 2, pp. 128-137, February 1987.

[8] P. Papazian, "Basic Transmission Loss and Delay Spread Measurements for Frequencies Between 430 and 5750 MHz," *IEEE Trans. Ant. Prop.*, vol. 53, no. 2, pp. 694-701, February 2005.

[9] J. R. Hampton, N. M. Merheb, W. L. Lain, D. E. Paunil, R. M. Shuford, W. T. Kasch, "Urban Propagation Measurements for Ground Based Communication in the Military UHF Band," *IEEE Trans. Ant. Prop.*, vol. 54, no. 2, pp. 644-654, February 2006.

[10] R. J. C. Bultitude, T. C. W. Schenk, N. A. A. Op den Kamp, N. Adnani, "A Propagation-Measurement-Based Evaluation of Channel Characteristics and Models Pertinent to the Expansion of Mobile Radio Systems to Frequencies Beyond 2 GHz," *IEEE Trans. Veh. Tech.*, vol. 56, no. 2, pp. 382-388, March 2007.

[11] D. W. Matolak, "Channel Modeling for Vehicle-to-Vehicle Communications," *IEEE Communications Magazine,* vol. 46, no. 5, pp. 76-83, May 2008.

[12] C. Gentile, N. Golmie, K.A. Remley, C.L. Holloway, and W.F. Young, "A channel propagation model for the 700 MHz band," *Proc. 2010 IEEE Int. Conf. on Comm. (ICC),* 2010, pp. 1 – 6.

[13] W.F. Young, C.L. Holloway, G. Koepke, D. Camell, Y. Becquet, and K.A. Remley, "Radio-wave propagation Into large building structures—part 1: CW signal attenuation and variability, *IEEE Trans. Antennas Propagat.*, vol. 58, no. 4, Apr. 2010, pp. 1279-1289.

[14] K. A. Remley, G. Koepke, C. L. Holloway, C. Grosvenor, D. Camell, J. Ladbury, R. T. Johnk, and W. F. Young, "Radio Wave Propagation Into Large Building Structures; Part 2, Characterization of Multipath," *IEEE Trans. Antennas Propagat.*, vol. 58, no. 4, Apr. 2010, pp. 1290-1301.

[15] W. F. Young, K. A. Remley, J. Ladbury, C. L. Holloway, C. Grosvenor, G. Koepke, D. Camell, S. Floris, W. Numan, and A. Garuti, "Measurements to support public safety communications: attenuation and variability of 750 MHz radio wave signals in four large building structures," *NIST Technical Note 1552*, Aug. 2009.

[16] W. F. Young, K. A. Remley, D. W. Matolak, Q. Zhang, C. L. Holloway, C. Grosvenor, C. Gentile, G. Koepke, and Q. Wu, "Measurements and models for the wireless channel in a ground-based urban setting in two public safety frequency bands," *NIST Technical Note 1557*, Jan. 2011.

[17] J.C.-I. Chuang, "The effects of time delay spread on portable radio communications channels with digital modulation," *IEEE J. Selected Areas in Comm.*, vol. SAC-5(5), pp. 879-889, June 1987.

[18] Y. Oda, R. Tsuchihashi, K. Tsuenekawa, M. Hata, "Measured path loss and multipath propagation characteristics in UHF and microwave frequency bands for urban mobile communications," *Proc. 53rd IEEE Vehic. Technol. Conf.*, vol. 1, pp. 337-341, May 2001.

[19] J.A. Wepman, J.R. Hoffman, L.H. Loew, "Impulse Response Measurements in the 1850-1990 MHz Band in Large Outdoor Cells", *NTIA Report 94-309*, June 1994.

[20] B. N. Taylor, C. E. Kuyatt, "Guidelines for Evaluating and Expressing the Uncertainty of NIST Measurement Results," *NIST Technical Note 1297*, September, 1994.

[21]   J. B. Coder, J. M. Ladbury, C. L. Holloway, and K. A. Remley, "Examining the True Effectiveness of Loading a Reverberation Chamber: How to Get Your Chamber Consistently Loaded." *IEEE Int. Symp. Electromagnetic Compat.*, July 25-30, 2010.

Appendix A: Measured Data

A.1 Denver Urban Canyon

Location and Notes	Test Point	VNA Loss Data (dB)	Path Loss @750 MHz (mean of 9) (dB)	RMS Delay Spread @750 MHz (mean of 9) (ns)	908 MHz PASS Unit 1	908 MHz PASS Unit 2	450 MHz PASS
Denver Urban Canyon TX1	1	17.25	59.2	51.7	O	O	O
	2	5.15	44.1	20.6	O	O	O
	5	21.01	63.0	107.3	O	O	O
	7	34.24	76.2	146.1	O	O	O
	9	36.54	78.5	134.7	O	X	O
	10	43.48	85.5	211.0	O	O	O
TX2	1	36.84	78.8	114.3	O	O	O
	2	17.06	59.0	32.5	D	O	O
	5	27.65	69.6	81.6	O	X	O
	7	40.13	82.1	108.5	O	O	O
	9	42.34	84.3	94.8	O	X	O
	10	49.71	91.7	N/A	O	O	O

O = Alarm received
D = Alarm received with Delay
X = Alarm not received

Measurement details
Calibration Attenuator: 40 dB for the high bands, 20 dB for the low bands
Note: Denver Urban Canyon data: attenuator in place during measurements. No path loss correction for atten needed.
Reference measurement distance: 4 m (ref. from OATS => 20 dB atten)
Formula: VNA data + free-space loss = Path Loss

A.2 Horizon West Apartment Building

Location and Notes	Test Point	VNA Loss Data (dB)	Path Loss @750 MHz (dB)	RMS Delay Spread @750 MHz (ns)	908 MHz PASS	908 MHz PASS w/ repeater	450 MHz PASS
Horizon West Floor 2 Notes: - System 1 repeater at test point 2.	1	24.75	76.7	25.0	D	X	X
	2	21.41	73.4	25.0	D	X	X
	3	16.29	68.3	16.1	D	D	X
	4	16.97	69.0	32.1	D	X	X
	5	29.74	81.7	18.3	D	X	X
	6	35.44	87.4	54.3	X	X	X
	7	26.29	78.3	37.5	D	X	X
	8	26.34	78.3	48.5	X	X	X
	9	24.05	76.0	19.7	D	X	X
	10	24.25	76.2	46.8	X	X	X
	11	22.38	74.4	31.2	X	X	X
	12	20.68	72.7	20.1	D	X	X
	13	25.77	77.8	34.8	X	X	X
Horizon West Floor 7	1	28.22	80.2	33.9	D	X	O
	2	18.68	70.7	24.4	D	X	O
	3	17.16	69.1	29.8	D	X	O
	4	35.12	87.1	51.4	D	X	O
	5	37.71	89.7	52.4	D	X	O
	6	33.78	85.8	77.1	D	X	X
	7	34.08	86.1	46.2	D	X	O
	8	30.62	82.6	41.5	O	X	X
	9	28.92	80.9	51.3	D	X	O
	10	31.09	83.1	56.5	D	X	X
	11	33.51	85.5	47.4	D	X	X
	12	36.38	88.4	69.7	D	X	O
	13	28.50	80.5	28.2	D	X	O

O = Alarm received
D = Alarm received with Delay
X = Alarm not received
Measurement Details
Calibration Attenuator: 30 dB for the high bands, 30 dB for the low bands
Reference measurement distance: 4 m (OATS => 20 dB atten)
Note: 2 m ref file (collected at Horizon West) gives same PL within 0.01 dB.
Formula: VNA data + atten – ref meas atten + free-space loss = Path Loss

A.3 Republic Plaza Office Building

Location and Notes	Test Point	VNA Loss Data (dB)	Path Loss @750 MHz (dB)	RMS Delay Spread @750 MHz (ns)	908 MHz PASS	908 MHz PASS w/ repeater	450 MHz PASS
Republic Plaza Notes: - System 1 repeater at test point 2.	1	7.23	69.2	45.0	O	O	O
	2	27.06	89.0	39.5	D	O	O
	3	38.15	100.1	52.3	X	X	O
	4	37.60	99.6	133.4	X	X	O
	5	37.18	99.2	81.3	X	O	O
	6	42.26	104.2	102.8	X	O	O
	7	46.04	108.0	138.3	X	X	O
	8	44.88	106.9	104.7	X	O	O
	9	48.30	110.3	376.1	X	X	O
	10	45.34	107.3	338.2	X	O	O
	11	50.25	112.2	167.9	X	O	O
	12	50.48	112.5	231.6	X	O	O
	13	50.98	113.0	209.1	X	O	O
	14	51.82	113.8	192.3	X	O	O
	15	49.60	111.6	240.2	X	O	O
	16	44.64	106.6	377.5	X	X	O
	17	29.28	91.3	296.9	O	O	O
	18	30.45	92.4	161.8	O	O	O
	19	42.24	104.2	429.9	O	O	O
	20	39.30	101.3	333.3	O	O	O
	21	47.07	109.1	453.5	O	O	O

O = Alarm received
D = Alarm received with Delay
X = Alarm not received

Measurement Details
Calibration attenuator (atten): 30 dB for the high bands, 10 dB for the low bands
Reference measurement distance (free-space loss): 4 m (OATS => 20 dB atten used)
Formula: VNA data + atten + free-space loss = Path Loss

A.4 NIST Building 27

Location and Notes	Test Point	VNA Loss Data (dB)	Path Loss @750 MHz (dB)	RMS Delay Spread @750 MHz (ns)	908 MHz PASS	908 MHz PASS w/ repeater	450 MHz PASS
Building 27 Notes: - No PASS tests conducted.	1LOS	47.26	109.2	228.2	--	--	--
	2LOS	30.18	92.2	18.4	--	--	--
	3LOS	22.89	84.9	13.4	--	--	--
	1NLOS	48.43	110.4	--	--	--	--
	2NLOS	35.61	97.6	15.9	--	--	--
	3NLOS	21.17	90.2	11.7	--	--	--

Measurement details

Calibration Attenuator: 40 dB for the high bands, 20 dB for the low bands
Reference measurement distance: 4 m (ref. from OATS => 20 dB atten)
Formula: VNA data + atten + free-space loss = Path Loss

A.5 NIST Building 24

Outside-to-Inside

Location and Notes	Test Point	VNA Loss Data (dB)	Path Loss @750 MHz (mean of 9) (dB)	RMS Delay Spread @750 MHz (mean of 9) (ns)	908 MHz PASS	908 MHz PASS w/ repeater	450 MHz PASS
NIST Bldg 24 Notes: - Path Loss and RMS Delay calculated with TX outdoors. - System 1 repeater at test point 2.	1	37.65	99.63	39.5	O	O	O
	2	46.55	108.5	N/A	O	O	O
	3	53.16	115.1	N/A	O	O	O
	4	--	--	N/A	X	O	O
	5	--	--	N/A	X	D	X
	6	--	--	N/A	X	D	X
	7	--	--	N/A	X	O	X
	8	52.81	114.8	N/A	D	D	X
	9	51.90	113.9	N/A	X	D	X
	10	--	--	N/A	X	O	X

O = Alarm received
D = Alarm received with Delay
X = Alarm not received

Measurement details
Calibration attenuator: 40 dB for the high bands, 20 dB for the low bands
Ref measurement distance: 4 m (OATS => 20 dB atten)
Formula: VNA data + atten + free-space loss = Path Loss

A.6 NIST Building 1 Corridor

Outside-to-Inside

Location and Notes	Test Point	VNA Loss Data (dB)	Path Loss @750 MHz (dB)	RMS Delay Spread @750 MHz (ns)	908 MHz PASS	908 MHz PASS w/ repeater	450 MHz PASS
NIST Bldg 1 Corridor: Inside-Outside Notes: - Base outside at "Wing 6" - 908 MHz repeater at test point 8.	1	49.7	105.7	55.9	O	O	O
	2	56.2	112.2	53.8	O	O	O
	3	60.3	116.3	34.1	O	O	O
	4	70.1	126.0	95.1	O	O	O
	5	72.3	128.3	88.1	O	O	O
	6	75.1	131.1	84.0	O	O	D
	7	73.9	129.9	79.1	O	O	D
	8	77.8	133.7	72.4	D	O	X
	9	74.1	120.1	36.9	X	D	X
	10	84.2	140.2	N/A	X	O	X
	11	86.4	142.4	N/A	X	O	X
	12	85.2	141.1	N/A	X	D	X
	13	87.9	143.9	N/A	X	D	X
	14	86.5	142.4	N/A	X	O	X
	15				X	O	X
	16				O	O	X
	17				X	O	X
	18				X	O	X

O = Alarm received
D = Alarm received with Delay
X = Alarm not received

Measurement Details
Calibration attenuator: 40 dB for the high bands, 20 dB for the low bands
Reference measurement distance: 2 m
Formula: VNA data + atten + free-space loss = Path Loss

A.6 (continued) NIST Building 1 Corridor

Inside-to-Inside

Location and Notes	Test Point	VNA Loss Data (dB)	Path Loss @750 MHz (dB)	RMS Delay Spread @750 MHz (ns)	908 MHz PASS	908 MHz PASS w/ repeater	450 MHz PASS
NIST Bldg 1 Corridor: Inside-Inside	1	10.36	66.3	11.8	O	O	O
	2	28.46	84.4	25.5	O	O	O
	3	39.16	95.1	27.1	O	O	O
	4	45.01	101.0	28.4	O	O	O
Notes:	5	50.19	106.2	28.5	O	O	O
-TX located	6	60.40	116.4	36.0	O	O	O
on "Wing 4"	7	65.22	121.2	57.6	O	O	O
loading	8	67.91	123.9	30.9	O	O	O
dock	9	--	--	--	O	O	O
- 908 MHz	10	--	--	--	O	O	X
repeater at	11	--	--	--	X	O	X
test point 4.	12	--	--	--	X	O	X
- Path loss	13	--	--	--	X	O	X
tests not	14	--	--	--	X	O	X
possible at	15	--	--	--	X	O	X
points 9-15.	16	81.39	137.4	62.54	D	O	X
	17	91.54	147.5	43.10	X	O	X
	18	94.83	150.8	N/A	X	O	X

O = Alarm received
D = Alarm received with Delay
X = Alarm not received

Measurement Details
Calibration attenuator: 40 dB for the high bands, 20 dB for the low bands
Reference measurement distance: 2 m
Formula: VNA data + atten + free-space loss = Path Loss

A.7 Colorado Convention Center

Location and Notes	Test Point	VNA Loss Data (dB)	Path Loss @1 GHz (dB)	RMS Delay Spread @ 1 GHz (ns)	908 MHz PASS	908 MHz PASS w/ repeater	450 MHz PASS
Colorado Convention Center Notes: - 450 MHz system tested only.	1	24.65	96.6	85.6	--	--	O
	2	26.92	98.9	68.2	--	--	O
	3	37.47	109.5	79.6	--	--	O
	4	55.00	127.0	159.3	--	--	O
	5	73.05	145.0	180.6	--	--	D
	6	82.61	154.6	128.8	--	--	X

O = Alarm received

D = Alarm received with Delay

X = Alarm not received

Measurement details

Note: Data shown are for 1 to 1.2 GHz band, horn antenna to omnidirectional antenna.
Path loss shown is higher than expected for PASS performance due to antenna gain.
Calibration attenuator: 30 dB for the high band
Reference measurement distance: 3 m
Formula: VNA data + atten + free-space loss = Path Loss

Appendix B: RMS Delay Spread vs. Attenuation

B.1 Denver Urban Canyon
Transmit Site 1

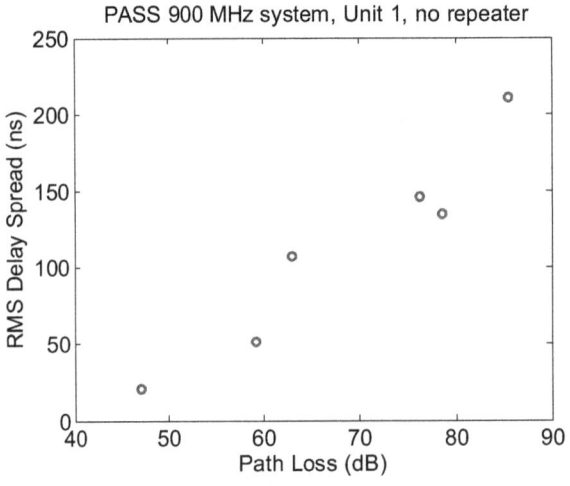

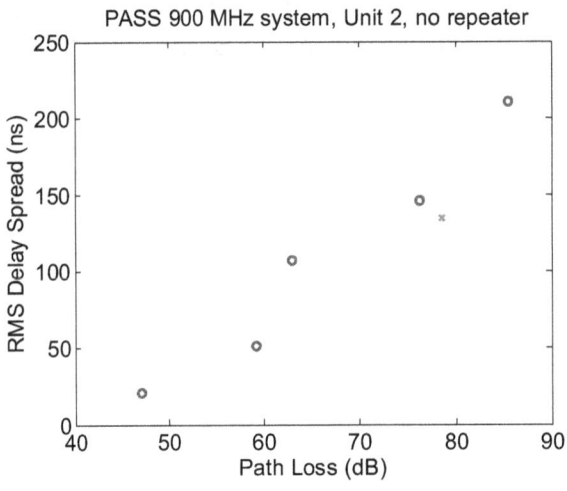

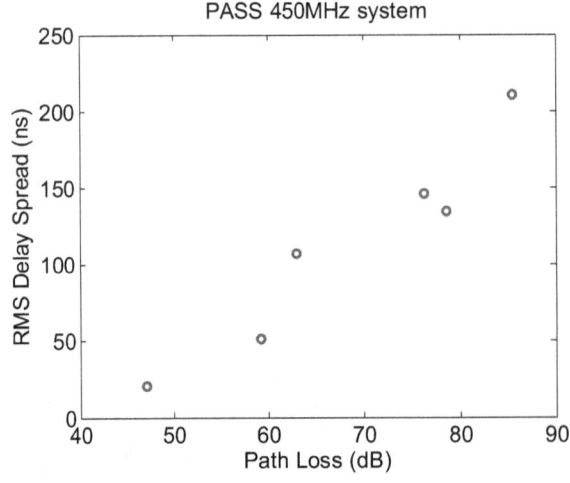

B.1 (continued) Denver Urban Canyon

Transmit Site 2

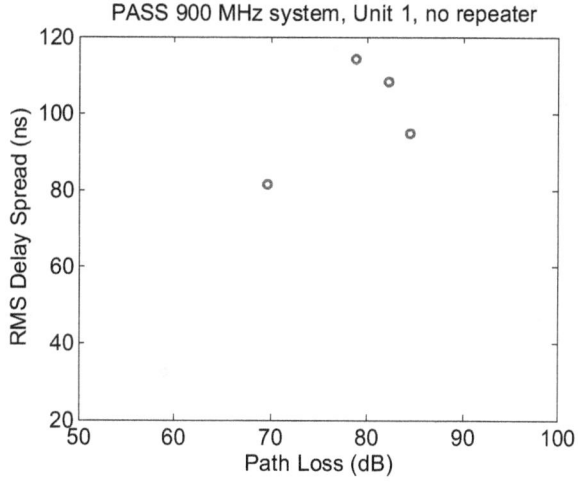

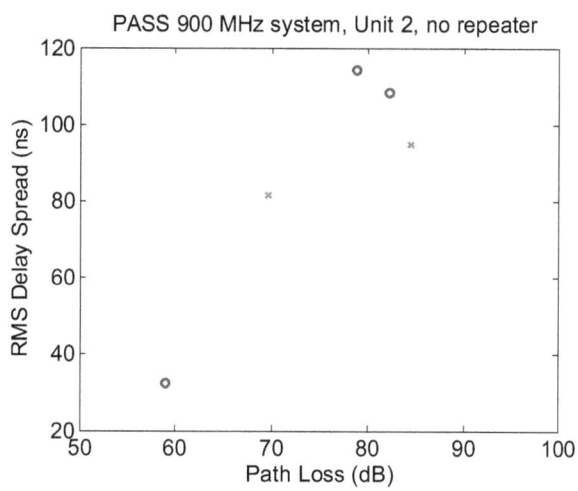

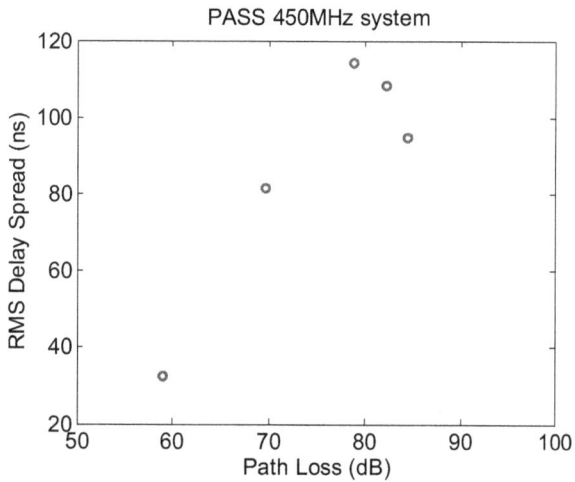

B.2 Horizon West Apartment Building
Floor 2

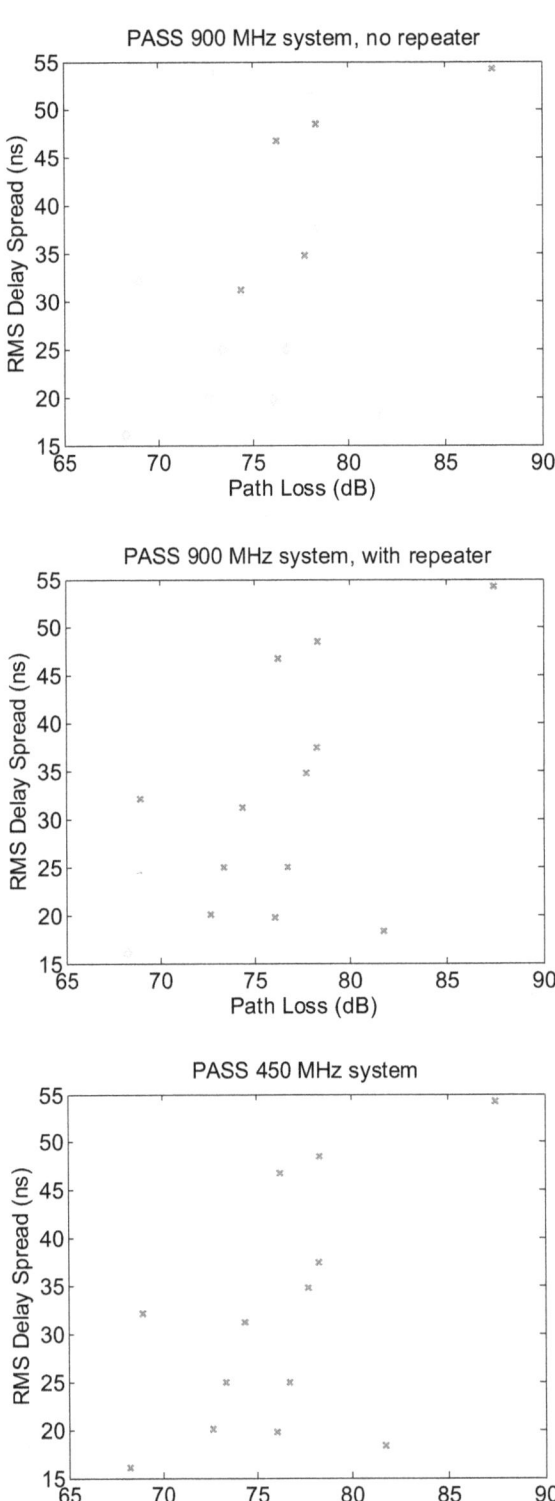

B.2 (continued) Horizon West Apartment Building

Floor 7

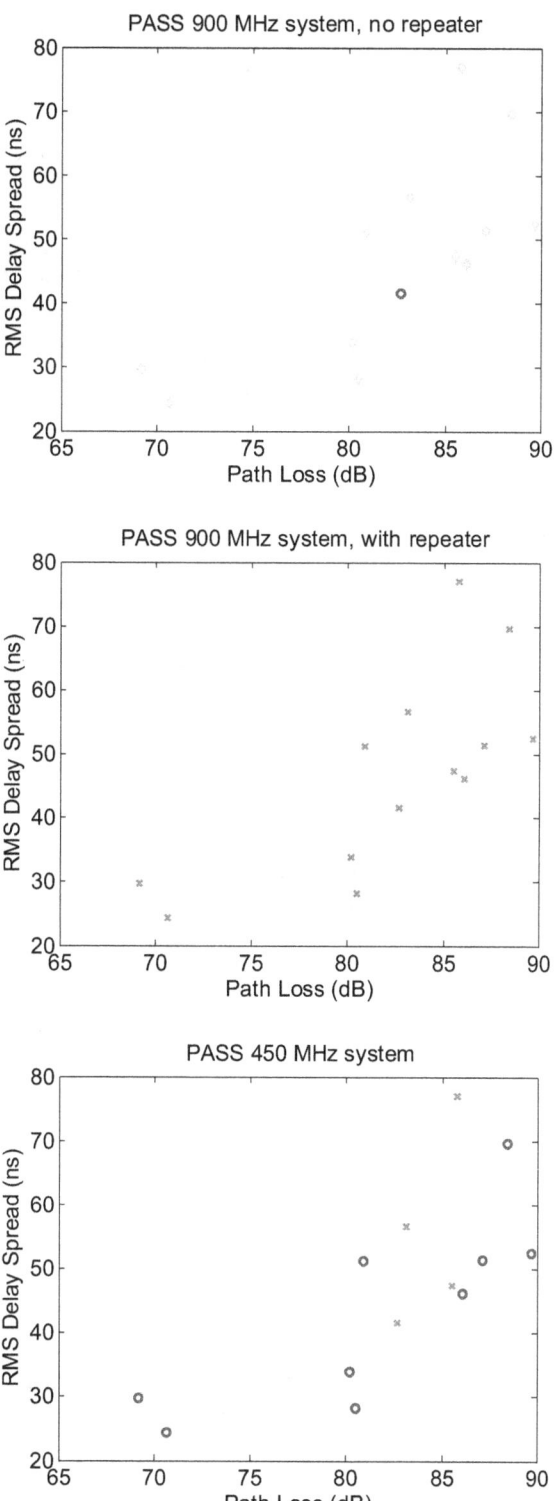

B.3 Republic Plaza Office Building

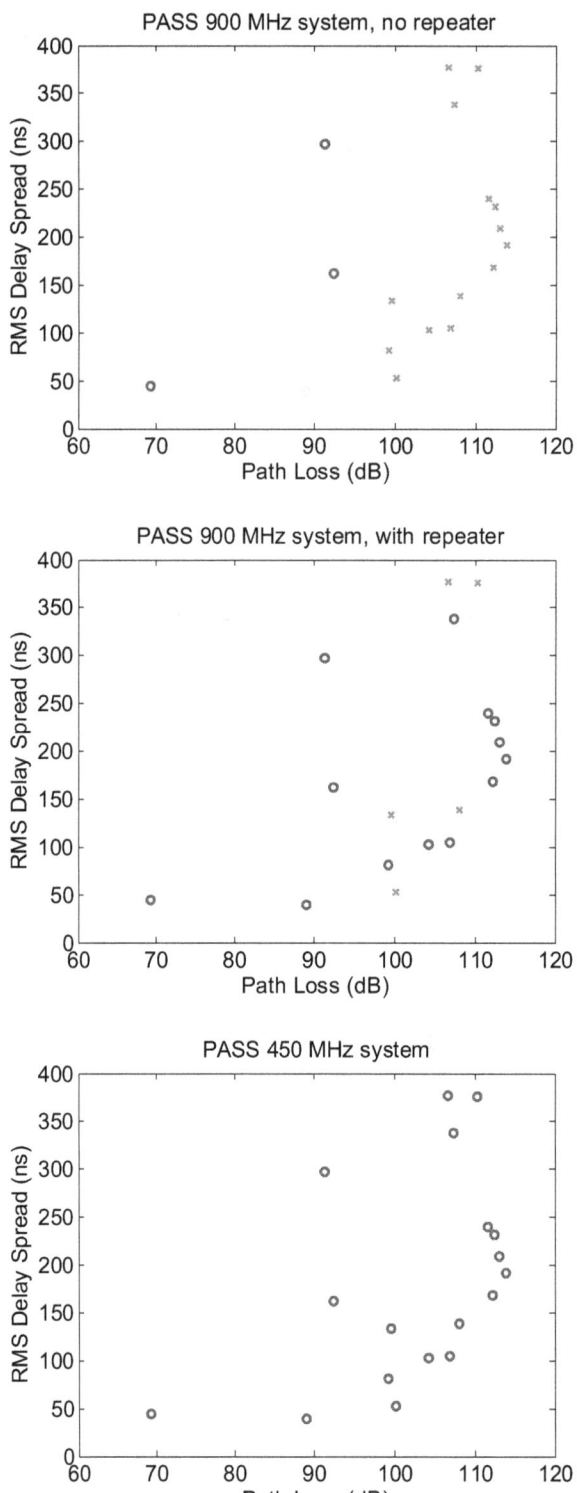

B.4 NIST Building 27

Small main building connected by long subterranean tunnel to small back room.

No PASS tests conducted, channel characterization only

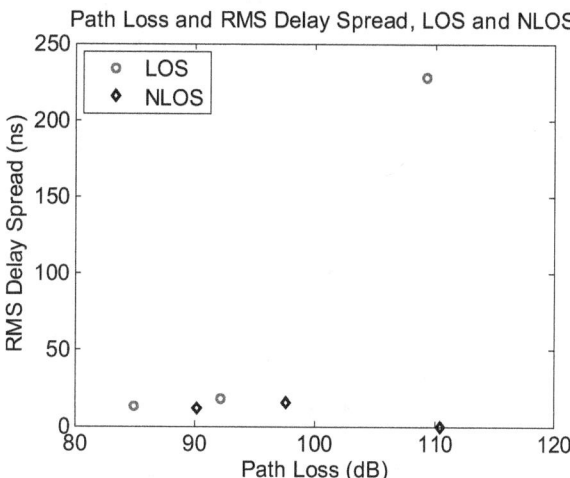

B.5 NIST Building 24

Outside-to-Inside

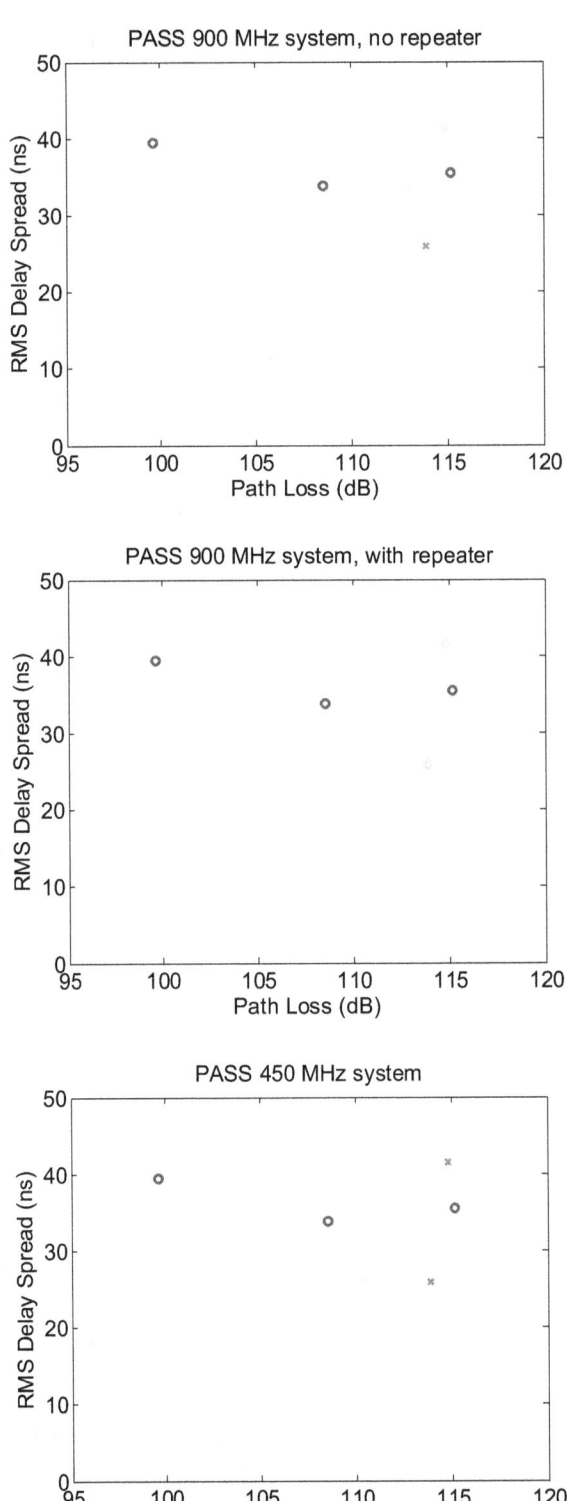

B.6 NIST Building 1 Corridor

Outside-to-Inside

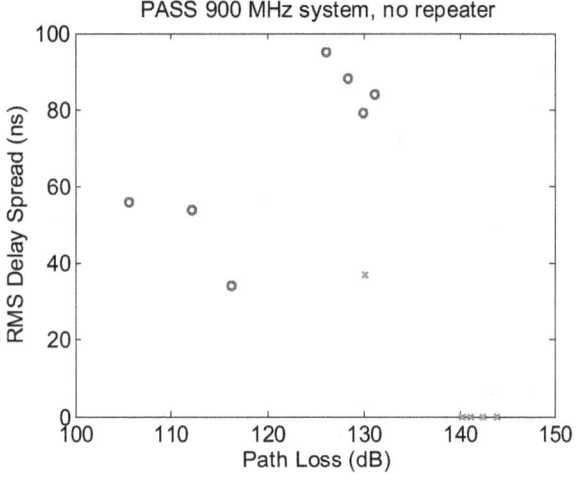

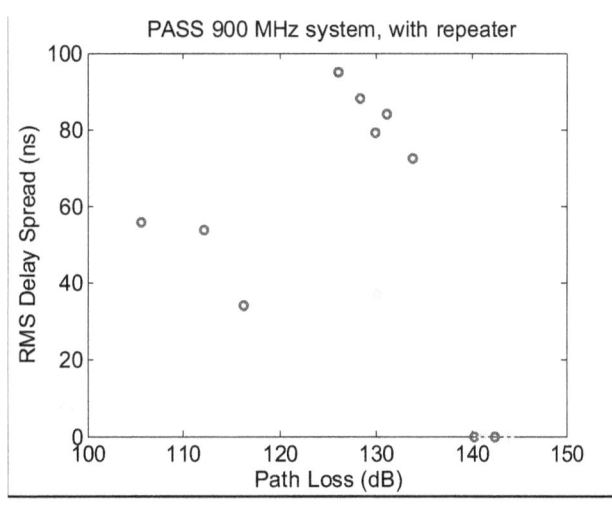

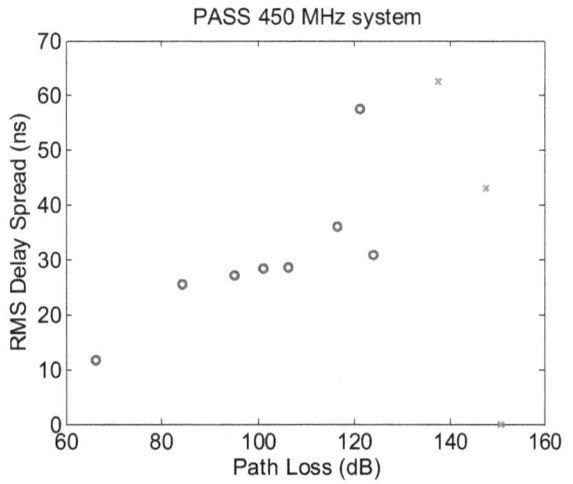

B.6 (continued) NIST Building 1 Corridor

Inside-to-Inside

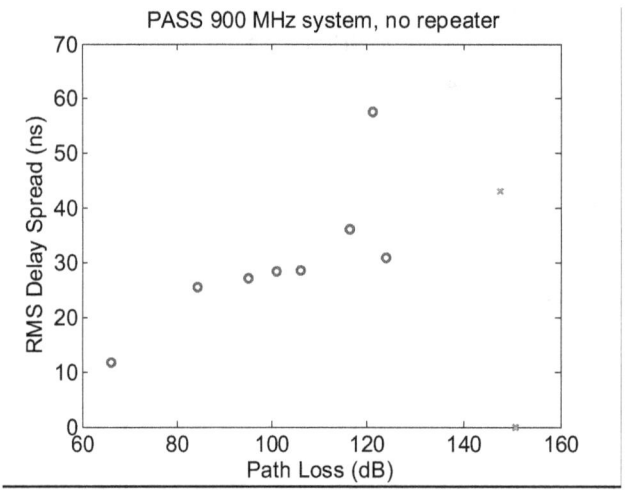

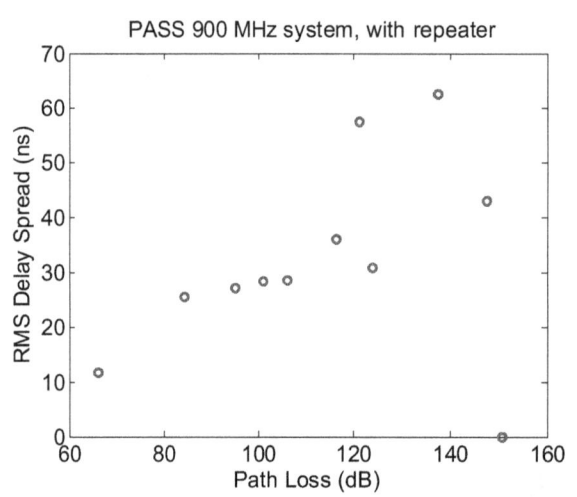

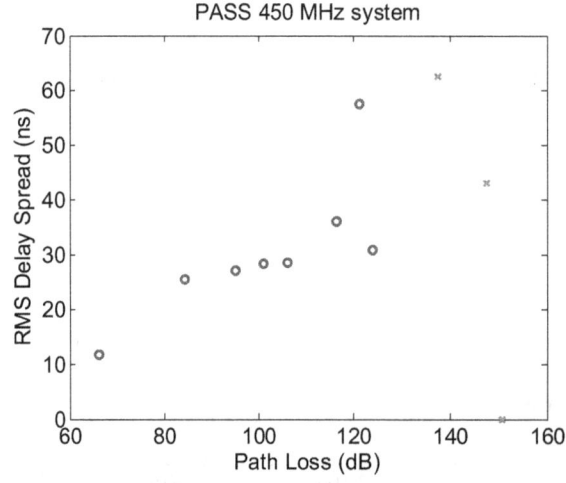

B.7 Colorado Convention Center

450 MHz PASS only

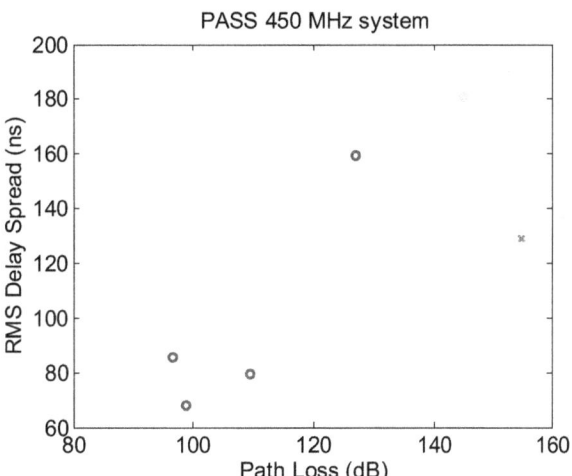

Appendix C: Path Loss

Free space path loss:

$$PL = 10*\log_{10}[(\lambda/4\pi d)^2]\ (dB),$$

where

$\lambda = c/f$,

f=carrier frequency (Hz)

c= speed of light, 3e8 m/s

At 750 MHz:

2 m: PL = 35.96 dB

3 m: PL = 39.49 dB

4 m: PL = 41.98 dB

NIST *Technical Publications*

Periodical

Journal of Research of the National Institute of Standards and Technology: Reports NIST research and development in metrology and related fields of physical science, engineering, applied mathematics, statistics, biotechnology, and information technology. Papers cover a broad range of subjects, with major emphasis on measurement methodology and the basic technology underlying standardization. Also included from time to time are survey articles on topics closely related to the Institute's technical and scientific programs. Issued six times a year.

Nonperiodicals

Monographs: Major contributions to the technical literature on various subjects related to the Institute's scientific and technical activities.

Handbooks: Recommended codes of engineering and industrial practice (including safety codes) developed in cooperation with interested industries, professional organizations, and regulatory bodies.

Special Publications: Include proceedings of conferences sponsored by NIST, NIST annual reports, and other special publications appropriate to this grouping such as wall charts, pocket cards, and bibliographies.

National Standard Reference Data Series: Provides quantitative data on the physical and chemical properties of materials, compiled from the world's literature and critically evaluated. Developed under a worldwide program coordinated by NIST under the authority of the National Standard Data Act (Public Law 90-396). NOTE: The Journal of Physical and Chemical Reference Data (JPCRD) is published bimonthly for NIST by the American Institute of Physics (AlP). Subscription orders and renewals are available from AIP, P.O. Box 503284, St. Louis, MO 63150-3284.

Building Science Series: Disseminates technical information developed at the Institute on building materials, components, systems, and whole structures. The series presents research results, test methods, and performance criteria related to the structural and environmental functions and the durability and safety characteristics of building elements and systems.

Technical Notes: Studies or reports which are complete in themselves but restrictive in their treatment of a subject. Analogous to monographs but not so comprehensive in scope or definitive in treatment of the subject area. Often serve as a vehicle for final reports of work performed at NIST under the sponsorship of other government agencies.

Voluntary Product Standards: Developed under procedures published by the Department of Commerce in Part 10, Title 15, of the Code of Federal Regulations. The standards establish nationally recognized requirements for products, and provide all concerned interests with a basis for common understanding of the characteristics of the products. NIST administers this program in support of the efforts of private-sector standardizing organizations.

*Order the **following** NIST publications: FIPS and NISTIRs: from the National Technical Information Service, Springfield, VA 22161.*

Federal Information Processing Standards Publications (FIPS PUB): Publications in this series collectively constitute the Federal Information Processing Standards Register. The Register serves as the official source of information in the Federal Government regarding standards issued by NIST pursuant to the Federal Property and Administrative Services Act of 1949 as amended, Public Law 89-306 (79 Stat. 1127), and as implemented by Executive Order 11717 (38 FR 12315, dated May 11,1973) and Part 6 of Title 15 CFR (Code of Federal Regulations).

NIST Interagency or Internal Reports (NISTIR): The series includes interim or final reports on work performed by NIST for outside sponsors (both government and nongovernment). In general, initial distribution is handled by the sponsor; public distribution is handled by sales through the National Technical Information Service, Springfield, VA 22161, in hard copy, electronic media, or microfiche form. NISTIRs may also report results of NIST projects of transitory or limited interest, including those that will be published subsequently in more comprehensive form.

U.S. Department of Commerce
National Institute of Standards and Technology
325 Broadway
Boulder, CO 80305-3337

Official Business
Penalty for Private Use $300

_____